WATERWORKS: Improving Irrigation Management in Mexican Agriculture

Ronald G. Cummings
Victor Brajer
James W. McFarland
Jose Trava
Mohamed T. El-Ashry

WORLD RESOURCES INSTITUTE

WRI Paper #5
December 1989

Library of Congress Cataloging-in-Publication Data

Waterworks: Improving irrigation management in Mexican agriculture /
by Ronald G. Cummings . . . [et al.] : with an appendix by Manuel Puebla.
p. cm.—(World Resources Institute report)
Bibliography: p.
ISBN 0-915825-33-3 : $10.00
1. Food supply—Government policy—Mexico. 2. Water resources development—Mexico. 3. Irrigation farming—Government policy—Mexico. 4. Agriculture and state—Mexico—History—20th century.
I. Cummings, Ronald G. II. World Resources Institute. III. Series.
HD9014.M62I46 1989
333.91′3′0972—dc20 89-9041
CIP

Kathleen Courrier
Publications Director

Brooks Clapp
Marketing Manager

Hyacinth Billings
Production Manager

Each World Resources Institute Report represents a timely, scientific treatment of a subject of public concern. WRI takes responsibility for choosing the study topics and guaranteeing its authors and researchers freedom of inquiry. It also solicits and responds to the guidance of advisory panels and expert reviewers. Unless otherwise stated, however, all the interpretation and findings set forth in WRI publications are those of the authors.

CONTENTS

ACKNOWLEDGMENTS

In the course of this study, many individuals provided helpful guidance, information, and constructive comments. The authors extend their appreciation and gratitude to all of them, but wish to single out Arturo Gomez Pompa, William Easter, Gerald O'Mara, Aaron Zazueta, Gus Speth, and Robert Repetto, who reviewed an earlier draft of the manuscript. The authors also wish to thank those who assisted in the research process—Melissa Paly and Melissa Wing.

A number of individuals in Mexico were particularly generous with data and with insights into the contemporary problems of Mexican agriculture. Chief among these are Dr. Donald Winklemann, Director, Centro Internacional por la Mejoramiento de Maiz y Trigo (CIMMYT); Dr. Francisco Oryzabal, SARH; and Dr. Javier Salas, Banco de Mexico.

Our thanks also to Kathleen Courrier for editing and sharpening the manuscript, Hyacinth Billings for producing the report, and Beverlyann V. Austin for general assistance.

R.C.
V.B.
J.M.
J.T.
M.E.

FOREWORD

Mexico today faces resource shortages that are likely to outlast its current indebtedness, inflation, and attendant efforts in rural development unless corrective action is taken. Before 1970, Mexico was a net exporter of such staples as beans, corn, and wheat. But by 1980, it was importing a quarter of its food supply, and in 1989 Mexico will spend more on food imports than ever. In addition, rapid population increases will continue to frustrate attempts at regaining food self-sufficiency.

Hard-pressed by population growth predicted to hover between 3.0 and 3.5 percent annually in the years ahead, Mexico has two basic options for expanding domestic food production, according to Dr. Cummings and the other authors of this report. It can bring more land under the plough, or it can increase agricultural output by using cultivated land and agricultural inputs—especially water—more efficiently and productively.

For reasons spelled out in *Waterworks: Improving Irrigation Management in Mexican Agriculture,* expanding total agricultural acreage is not the only answer. In general, little good arable land remains untilled, and the costs of moving on to less suitable lands are daunting in the best of times and probably out of the question now.

The second option has more than the process of elimination going for it. As noted in this report, history may be on Mexico's side in the struggle to wrest more food from every bit of agricultural land. The country made impressive productivity gains during the Green Revolution and again in 1980 and 1981—the early years of the Mexican Government's *Sistema Alimentario Mexicano,* a program designed to boost national self-sufficiency in basic foods and infuse the *campesino* economy with new vitality.

Historically, irrigated agriculture in Mexico has been far more productive than dryland farming, and a common thread in the productivity schemes that have succeeded is water management. By completing irrigation works left unfinished and better maintaining and managing existing systems, Mexico can increase agricultural yields. Other parts of the solution recommended here include better water conservation, stepped-up research on high-yielding crops that require little water, and improved agricultural extension programs that bring new technologies and techniques to more than the elite minority—the land-rich farmers who are already relatively prosperous.

Although based partly on precedents, this prescription will encounter institutional and political barriers that Dr. Cummings and his co-authors face squarely. Already the debt-burdened government has diverted irrigation-maintenance funds to complete projects halted in part by inflation. Land tenure reform remains far from adequate. And the rural poor are chronically underemployed and undereducated. Reversing these trends won't be easy. But, because the stakes are so high and the technological possibilities so great, policies like those spelled out here—policies that will minimize or surmount structural barriers and change spending priorities within the agricultural sector—are critically important.

Waterworks: Improving Irrigation Management in Mexican Agriculture complements other WRI research publications on water use and management. *Troubled Waters: New Policies for Managing Water in the American West* (Mohamed T. El-Ashry and Diana Gibbons, WRI, 1986) identifies creative policies for conserving, reallocating, and improving the efficiency of water supplies in the western United States. *Water and Arid Lands of the Western United States* (Mohamed T. El-Ashry and Diana Gibbons, co-editors, Cambridge University Press, 1988) bolsters the same arguments with detailed case studies of water use and water rights conflicts in Texas, Arizona, California, and Colorado. *Skimming the Water: Rent-Seeking and the Performance of Public Irrigation Systems* (Robert Repetto, WRI, 1986) shows how misguided economic incentives cost governments huge sums and distort investment decisions while inviting environmental abuse and wasting natural resources.

James Gustave Speth
President
World Resources Institute

I.

MEXICAN AGRICULTURE'S CHALLENGE

Agricultural production has figured centrally in Mexico's economic development plans. Historically, these plans have assumed the availability of cheap energy and of cheap food obtained partly through the use of highly subsidized agricultural inputs. Planners have also promoted self-sufficiency in beans and major grains.

"Mexican agriculture's failure to keep pace with the national consumption of agricultural products has changed Mexico from an exporter of major staples to an importer."

Unfortunately, both population growth and per capita consumption of the basic food crops—rice, beans, corn, and wheat—have increased more rapidly than agricultural production, and sustaining large subsidies for inputs has increasingly strained Mexico's economy. The annual rate of increase in Mexico's population has exceeded 3 percent over the last three decades (Alba, p. 752) and is expected to remain high for the rest of the century (Alba, 1982, pp. 756–758).[1] Meanwhile, the only basic food crop enjoying annual production increases of 3 percent or more in the 1970s was corn; production rates for rice, beans, and wheat, failed to keep up with population growth. *(See Table 1.)*

Mexican agriculture's failure to keep pace with the national consumption of agricultural products has changed Mexico from an exporter of major staples to an importer. *(See Table 2.)* In most years before 1970, Mexico was a net exporter of beans, corn, and wheat. During 1965, Mexico exported 1.3 million tons of corn—more than 15 percent of its total crop (SARH, September 1983, pp. 42–3). But Mexico began importing corn in the 1970s, and by the mid-1970s it had become a net importer of beans and wheat as well. In 1980, Mexico imported almost one fourth of its domestic consumption of corn and wheat and one third of its domestic consumption of beans (SARH, September 1983, pp. 34, 44, and 56). Imports fell to about 10 percent of domestic consumption in 1982 and 1983, but Mexico still imports more of these basic commodities than it exports.

Table 1. Trends in Mexico's Agricultural Sector, Selected Crops 1960–85

	Hectares (000)				**Production/Hectare (tons)**			
Crop	**1960**	**1970**	**1980**	**1985**	**1960**	**1970**	**1980**	**1985**
Rice	143	150	124	194	2,297	2,703	3,494	3,077
Beans	1,326	1,747	1,551	2,166	398	530	603	610
Corn	5,558	7,440	6,776	7,500	975	1,194	1,828	1,673
Wheat	840	886	724	1,100	1,417	3,020	3,848	4,128
Cartamo	26	175	416	367	1,248	1,645	1,152	962
Soybeans	4	112	154	440	1,230	1,920	2,092	1,889
Sorghum	165	940	1,590	1,568	1,797	2,829	3,039	3,520
Sugar	346	547	546	482	56,432	63,381	66,875	72,800
Cebada	240	224	319	321	752	1,060	1,659	1,819
Cotton	900	400	355	275	523	812	1,050	1,040

	Total Production (million tons)			
Crop	**1960**	**1970**	**1980**	**1985**
Rice	328	405	445	597
Beans	528	925	935	1,322
Corn	5,419	8,879	12,374	12,545
Wheat	1,190	2,676	2,785	4,541
Cartamo	32	289	480	353
Soybeans	5	215	322	831
Sorghum	209	2,747	4,689	5,520
Sugar	19,542	34,651	36,480	35,094
Cebada	180	238	530	584
Cotton	470	333	373	286

Source: Secretaria de Agricultura y Recursos Hidraulicos, *Econotecnia Agricola,* "Consumos Aparentes de Productos Agricolas, 1925–82," Vol. VII, No. 9, September 1983, Mexico, D.F. For 1985, see Secretaria de Agricultura y Recursos Hidraulicos, *Programa Nacional de Desarrollo Rural Integral 1985–88,* Mexico, D.F., 1985, p. 103.

The declining terms of trade in the agricultural sector during the late 1970s and early 1980s no doubt hurt Mexican agriculture. Between 1980 and 1984, producer prices for agricultural products (cereals and grains) in general increased by 764 percent, but farm prices for rice, corn, and wheat increased by only 225 percent and bean prices by 300 percent, largely because of government-imposed price controls (SARH, 1981, p. 264). Moreover, prices for pesticides, petroleum products, and agricultural tractors and other farm inputs increased by 852 percent, 1,788 percent, and 913 percent, respectively (Banco de Mexico, 1984).

Even though overall per capita consumption of basic foods increased during this period of declining terms of trade, the diets of Mexico's rural poor (an estimated 40 percent of the rural population) deteriorated (Barkin, 1983, pp. 13–19). Indeed, more than 35 million people in rural areas now suffer severe calorie and protein deficiencies (Aamorano U., 1985; *see* Chonchol, 1982, pp. 253–65).[2] Since

Table 2. Per Capita Consumption and Net Imports of Basic Crops in Mexico, 1960–80 and 1988 "Goals"

Crop	1960	1965	1970	1975	1980	1988*	
			A. Per Capita Consumption in Rural Mexico of: (Kilograms/person/year)				
Rice	6.56	6.25	5.60	7.90	5.60	7.20	(04.10**)
Beans	15.34	19.76	18.53	18.89	19.84	17.70	(18.50**)
Corn	138.46	178.06	190.13	185.33	238.81	161.40	(165.50**)
Wheat	33.13	34.62	51.99	47.43	53.12	33.40	(26.00**)
			B. Net Imports of: (10^3 tons)				
Rice	22.3	17.8	16.3	0	95.0		
Beans	24.7	−16.1	−61.2	104.2	441.0		
Corn	−429.0	−1335.0	758.0	2654.0	4187.0		
Wheat	4.2	−672.0	−40.0	43.0	900.0		

Source: Secretaria de Agricultura y Recursos Hidraulicos (SARH), *Econotecnia Agricola,* "Consumos Aparentes de Productos Agricolas, 1925–82, Vol. VII, No. 9, September 1983, Mexico, D.F.

*SARH, *Programa Nacional de Desarrollo Rural Integral 1985–88,* Mexico, D.F., 1985, p. 98.
**"Goals" are for the total population; numbers given in parentheses are goals for the rural poor. Instances where goals for rural populations exceed those for the total population reflect differences in the composition of diets, with the rural poor being more reliant on beans and corn.

1980, the Mexican government has experimented with one diet-enhancement program for the rural poor (the Mexican Food System—SAM), discarded it, and in 1985 instituted a second program (the National Program for Integrated Rural Development; hereafter called the National Program). If the new program is to succeed, agricultural production must surpass 1985 levels. *(See Table 3.)*

Will Mexican agriculture be able to meet the growing demand for agricultural products? To simply keep pace with expected population growth and modestly improve the diets of the rural poor (see SARH 1985), Mexico must expand its agricultural base or increase agricultural productivity or both. *(See also Nacional Financiera,* #51, 1985, p. 1193.) If the National Program's 1988 agricultural output goals for just the six crop types listed in Table 3 were extended to 2000, irrigated acreage would have to be doubled and the amount of non-irrigated arable land under cultivation would have to be increased by 60 percent. Substantial increases in productivity would also be needed—most notably for rice, corn, and cartamo.

Mexico's greatest hope for substantially boosting agricultural production most likely lies in increasing the productivity of farms—a realm where Mexican agriculture has historically been most successful. Although the productivity of rainfed agriculture deserves a great deal of government attention and support, this study focuses on the major inefficiencies in resource management plaguing Mexico's irrigated agriculture sector. It shows that improving irrigation could bring about badly needed increases in agricultural production. Predictably, most of these challenges involve inefficiencies in the management of the scarcest resource for irrigated farms: water.

Table 3. Mexico's Production Goals for 1988 Required for the "Integrated Rural Development Plan"

	Actual 1985		**1988 Goals**	
Crop	**Area Harvested (000 Ha.)**	**Yield/Ha. (Tons)**	**Area Harvested (000 Ha.)**	**Yield/Ha. (Tons)**
Basic Grains	10,960		11,709	
Corn	7,500	1.673	8,060	2.051
Wheat	1,100	4.128	1,145	4.162
Oilseed Crops	1,056		1,297	
Soybeans	440	1.889	520	1.967
Cartamo	367	.962	397	1.181
Fruits	444		461	
Hortalizas	115		121	
Forage Crops	1,828		2,277	
Sorghum	1,568	3.520	1,920	3.575
Industrial Crops	1,691		1,748	
Sugar	428	72.800	524	74.735
Cebada	321	1.819	329	1.915
Cotton	275	1.040	280	1.110
Coffee	420	.560	420	.579

Source: Secretaria de Agricultura y Recursos Hidraulicos, *Programa Nacional de Desarrollo Rural Integral 1985–88,* Mexico, D.F., 1985, p. 103.

IRRIGATION GEOGRAPHY

Mexico's recent productivity gains have been realized mainly on irrigated farms in northern Mexico. With few exceptions (notably, farms that produce sugar cane), irrigation works are concentrated in Baja California (North and South), Sonora, Sinaloa, Chihuahua, Durango, Coahuila, Nueva Leon, and Tamaulipas. Almost all of the irrigated production of beans, cartamo, and cotton takes place in the north, and 60 to 80 percent of irrigated production of rice, corn, and wheat does. *(See Table 4.)* Irrigation is extremely important in the production of six major crops: more than 70 percent of Mexico's rice, wheat, cartamo, and cotton, as well as 30 percent of its bean crops and 46 percent of sugar cane are grown on irrigated farms. As for corn, Mexico's most basic grain crop, irrigated farms comprising only 11 percent of Mexico's cultivated acreage in corn produce more than 20 percent of the nation's output.

Table 4. The Contribution of Total Irrigation and Irrigation in Northern States to Agricultural Production in Mexico, 1981

Percent of Total from Crop Irrigation	**Total Production in Mexico (000) Tons**	**Production from Irrigation (000) Tons**	**Percent of Total**	**Irrigated Production from Northern Mexico (000) Tons**
Rice 60	662	488	74	295
Beans 100	1,331	392	29	392
Corn 66	14,550	3,009	21	1,991
Wheat 83	3,192	2,972	93	2,459
Cartamo 100	339	242	71	242
Sugar 30	33,165	15,243	46	4,624
Cotton 99	973	939	97	934
Sorghum 51	6,086	2,615	43	1,332

Source: Secretaria de Agricultura y Recursos Hidraulicos, *Produccion Agricola Nacional, 1981 Anuario Estadistico,* Mexico, D.F., 1981.

THE PHYSICAL SETTING

Mexico, or more formally the United Mexican States, is divided into 29 states, two federal territories, and the federal district that contains the nation's capital, Mexico City. Mexico has a land surface area of approximately 762,000 square miles, making it close to one fifth the size of the United States and the third largest country in Latin America.[3] Its surface is made up of numerous mountains, volcanoes, and high plateaus, and small mountain ranges form many basins and valleys. More than three fourths (77 percent) of Mexico's land area lies in semi-arid or arid zones, where more than half of the country's agricultural production takes place. Mexico's five major rivers—the Grijalva-Usumacinta, Papaloapan, Mexicalapa, Coatzacoalcos, and Panuco, which comprise more than half of the mean annual flow of *all* Mexican rivers—are all located in the tropical southeastern region (Cummings, 1972, pp. 1–3).

Since Mexico lies between approximately 15 degrees and 32 degrees north latitude and its surface area varies widely in altitude, the climate is very diverse, ranging from the dry steppe and desert terrain covering most of the northern half of the country to a rainy tropical zone in the southern half.

Rainfall is greatest in southwestern Mexico, where it averages from 40 to over 120 inches per year. In contrast, the North gets less than 16 inches per year. Most precipitation occurs in summer and early fall, and most regions have a rainy season and a dry season.

The area of major concern in this report includes the northern part of the central plateau, which extends from the northern border to south of Mexico City, as well as the Pacific lowlands to the west. These lowlands are very arid in the north, and agriculture would fail without extensive irrigation systems. Livestock production is important only in the drylands, where favorable range conditions exist.

The geographically uneven distribution of water resources in Mexico, combined with limits on supplies, has troubled all sectors of the Mexican economy. Eighty-five percent of the country's water resources lie in an area less than 500 meters high; yet, 70 percent of Mexico's population and 80 percent of the country's industry are located in areas above 500 meters. Roughly 40 percent of the country's total land surface lies in the northern and central plateaus at elevations above 1,300 meters; this area accounts for approximately half of Mexico's total agricultural output. Within this area, the arid states of Sonora and Sinaloa make up Mexico's major wheat-growing center, one of Mexico's most efficient agricultural areas (Cummings, 1972, pp. 2–4).

II.

THE CASE FOR INCREASING PRODUCTIVITY IN IRRIGATED AGRICULTURE

A basic fact of agriculture in Mexico is that its irrigated lands are far more productive than its non-irrigated lands. Of the 15 million hectares under cultivation by 1970, some 3.8 million hectares were irrigated. Productivity on those lands was double that on non-irrigated lands, and irrigated lands accounted for 40 percent of the value of Mexico's agricultural production. *(See Table 5.)* Between 1970 and 1981, that productivity gap widened. Irrigated lands expanded by 2.4 percent per year, compared with a 1.1-percent annual expansion in non-irrigated areas, while productivity increases in irrigation were almost *six times* the rates for non-irrigated areas. *(See Table 5.)* By 1981, the 30 percent of agricultural lands that were irrigated accounted for 50 percent of the value of Mexican agricultural production.

Unfortunately, attempts to realize irrigated agriculture's full potential in Mexico have run up against social and economic obstacles that the country has not been able to overcome, despite repeated efforts. The productivity gains realized so far have come at a price that Mexico is still trying to reckon with. Of the U.S. $4.2 billion the Mexican government spent on agriculture in 1980, 40 percent went to irrigation works, and however much economic sense it made to invest resources in the most productive farmers, doing so raised questions of social equity for the rural poor, most of whom work in dryland agriculture.

"By 1981, the 30 percent of agricultural lands that were irrigated accounted for 50 percent of the value of Mexican agricultural production."

This disparity, along with the general level of poverty and the growing need to import such basic foods as corn and beans, prompted Mexico's government to initiate the "Mexican Food System" (Sistema Alimentario Mexicano—SAM) in March 1980 to boost non-irrigated agriculture by better integrating it with the rest of the economy and by helping the poorest farmers increase their income. SAM's stated objectives were achieving national self-sufficiency in the production of basic foods, revitalizing the *campesino* economy to increase the supply of

Table 5. Changes in Area Harvested, Value of Production and Productivity in Mexico's Agricultural Sector, 1979–81

Item	1970	1981	Rate of Change
1. Area Harvested: (1,000 Ha.)			
Total	14,975	18,668	2.0%
Rainfed	11,133	13,422	1.7
Irrigated	3,842	5,246	2.9
2. Value of Production: (Billion 1986 pesos)			
Total	2.83	4.10	3.4%
Rainfed	1.68	2.08	2.0
Irrigated	1.15	2.00	5.2
3. Productivity: (Value/ha; thousand 1986 pesos)			
Total	188,982	219,623	1.4%
Rainfed	150,903	154,970	0.2
Irrigated	299,323	381,243	2.2

Source: World Bank, "Mexico Irrigation Subsector Survey—First Stage," Vol. II, Report No. 4516-ME, June 20, 1983, p. 84. 1960 pesos converted to 1986 pesos; see Table 27.

internally produced food and fiber, and improving the incomes and diets of roughly the poorest third of the country's population.

Through a combination of technological change (primarily the expanded use of fertilizers and improved seeds on non-irrigated farms); government subsidies of fertilizer (70 percent), seeds (25 percent), and credit (extended at interest rates of 3 to 12 percent when market rates were running well above 50 percent); and improved organization of production and distribution networks, SAM had some positive effects during its 33-month tenure. (Zamorano U., 1985). Agricultural production rose by 6 percent between 1980 and 1981—the highest rate of growth in two decades. 1981 proved to be a record year for grain production (28.6 million tons of ten basic grains) and a year in which Mexico saw food self-sufficiency in wheat, beans, and rice. By 1982, Mexico's grain imports had fallen from the 9 million to 10 million ton range in 1980 and 1981 to only 3.7 million tons. (Unusually heavy, well-timed rains in 1981 no doubt helped that year's harvest.)

Yet, SAM collapsed in 1983. A law enacted in 1982, *Ley de Fomento Agropecuario* (LFA) extended SAM to *all* farmers, including those on irrigated lands, seriously hurting the program's general income-redistribution goals. In addition, with the worsening fiscal crisis, federal spending was cut and SAM's rapid growth after the 1982 law passed came to be viewed as budget-breaking (*Nacional Financiera,* #52, 1985, p. 1211), especially in the "lame duck" last year of President Lopez Portillo's term (Zamorano U., 1985).

Faced with inflation running at about 100 percent, a deficit consuming almost 18 percent of GNP, and the possibility of forced default on external loan payments, the incoming President, Miguel de la Madrid, initiated a five-year "National Development Plan" *(Plan Nacional de Desarrollo)* to reduce inflation rates and budget deficits and to reassert government's control over exchange-rate policies. As an extension of this general national development plan, a five-year national food program (*Programa Nacional de Alimentacion,* PRONAL) was established in 1983. The program set specific goals for agricultural production and specified how these goals should be achieved. *(See Table 6.)* Specifically, the idea was to make basic foods affordable, construct "multiple service centers" for rural farmers, establish cooperative stores and rural warehouses to benefit 4 million rural farmers, increase scholarships for bilingual and bicultural education, and bring basic literacy skills to over 4 million persons (SARH, 1985, Chapters 6 and 8; see particularly pp. 126–28).

So far, this ambitious program has had high costs and little success (SARH, June 1985, p. 31). Indeed, in its attempt to improve the diets and living standards of the rural poor, PRONAL has replaced a subsidy to producers with a subsidy to consumers. (Zamorano U., 1985).

Larger economic forces at play have also thwarted efforts to make irrigated agriculture satisfy Mexico's needs. Since the early 1970s, Mexico's federal deficit has increased tremendously. In 1980, it amounted to 7 percent of GNP (compared with 2.8 percent in the United States); by 1982, the deficit amounted to 16.5 percent of GNP (compared with a U.S. deficit of 4.2 percent of GNP; see World Bank, 1983, Annex 3, p. 1 and U.S. Department of Commerce, 1986). As this deficit grew, public expenditures for building and keeping up irrigation systems were sharply reduced. Annual public construction expenditures, which had risen from an average of some $16 billion in 1970 to $60 billion in 1980, *declined* to about $45 billion in 1982 and 1983. Similarly, public expenditures for machinery and equipment, which increased from about $8 billion in 1970 to $28 billion in 1980 were down to $13 billion by 1982 and 1983 (*Instituto Nacional de Estadistica* 1983, 1985).

"Mexico's rapid inflation has hit agriculture harder than the rest of the economy and hit farmers on irrigated lands harder than their counterparts in rainfed areas."

To make matters worse, Mexico's rapid inflation has (as discussed earlier) hit agriculture harder than the rest of the economy and hit farmers on irrigated lands harder than their counterparts in rainfed areas.

Despite these enormous obstacles, investing in increased agricultural efficiency and productivity remains the key to meeting the agricultural production goals that Mexico has set for itself. Expanding acreage is clearly not the answer. Mexico's stated goal is to increase the amount of rainfed agricultural land under cultivation by some 60 percent and irrigated acreage by almost 100 percent between 1985 and 2000. *(See Table 6.)* But most of the country's irrigated production of rice, beans, wheat, cartamo, and cotton is in the North, where comparatively little irrigable land remains unused. Indeed, such land in the northern and central regions at a maximum totals some 1.9 million hectares, which is less than two thirds of the acreage required to meet Mexico's goals for the year 2000. Similarly, in the areas most suited to rainfed farming, south and central Mexico, less than 1.2 million uncultivated hectares remain. As Table 7 suggests, most lands available for substantially expanding irrigated agriculture are in the areas least suited for irrigation, and most of the non-irrigated lands so far unused are in the areas least suited to rainfed agriculture. Meeting the end-of-the-century production goals for non-irrigated agriculture strictly through acreage expansion would require developing virtually every hectare of arable non-irrigated land in Mexico.

Table 6. Meeting Mexico's Agricultural Production Goals for the Year 2000: Required Increases in Cultivated Area and Productivity on Irrigated and Non-Irrigated Lands

A. Cultivated Area (000 hectares)

	Area Cultivated: Non-Irrigated			**Area Cultivated: Irrigated**		
Crop	**1985**	**1988**	**2000**	**1981**	**1988**	**2000**
Rice	80	127	805	114	121	154
Beans	1,863	1,887	1,986	303	369	811
Corn	6,600	6,926	8,398	900	1,134	2,856
Wheat	253	265	319	847	880	2,856
Cartamo	154	154	154	213	234	246
Sorghum	1,051	1,398	4,371	517	522	542
Sugar	284	298	361	198	226	383
Cotton	19	19	19	256	280	401
Totals	10,304	11,074	16,413	3,348	3,766	6,418

Sources:

(i) For 1985 cultivated area for irrigation, irrigated acreage in each crop as a percent of total acreage in the crop during 1982, as reported in SARH, *Produccion Agricola Nacional, 1981 Anuario Estadistico,* Mexico, D.F., 1985, p. 103. Non-irrigated acreage is allocated the remaining acreage.

(ii) For 1988, the total increase in irrigated acreage, as reported for crop groups in *Programa Nacional de Desarrollo. . .,* p. 136, is allocated to individual crops according to percentages of each crop of its group as given in *ibid,* p. 103. The remaining 1985–88 increases, are allocated to non-irrigated lands.

(iii) Acreage to the year 2000 is extrapolated at the compound annual rate of increase in each crop's acreage between 1985 and 1988.

Physical limits aren't the only obstacles to expanding agricultural land. Economic limits are critical too, particularly the costs of irrigating new lands. As Table 8 shows, the initial investment costs of irrigation projects have generally increased as lands less suitable for irrigation were brought into production. In small-scale irrigation, which typically cannot realize the economies of scale obtainable in larger projects, per-hectare costs of new irrigated land increased from 338,000 pesos (1986 pesos) in 1947 to 2.3 million pesos three decades later. For large-scale projects, per-hectare costs increased from some 1.7 million pesos from 1947 through 1952 to some 2.7 million between 1977 and 1981. These data for large-scale projects are somewhat distorted since project-improvement and rehabilitation costs are included for 27 percent of the benefited areas in the first and second periods, 56 percent in the fourth period, and 52 percent in the fifth period. But per-hectare costs in the third and sixth periods, which include only investment costs, make it clear that the per-hectare costs are significantly higher in these periods than in any other. The highest per-hectare costs occur in the final period, which strongly suggests rising marginal costs for developing irrigated land. Calculated in terms of three-year moving averages for 1979 through 1982, the per-hectare costs of expanding irrigation areas with new large-scale irrigation projects rose from 1.7 million pesos in 1979 to 3.1 million pesos in 1982 alone. *(See Table 9.)*

Table 6. (Continued)

B. Agricultural Productivity (Tons per Hectare)

	Yields—Non-Irrigated			Yields—Irrigated		
Crop	**1985**	**1988**	**2000**	**1981**	**1988**	**2000**
Rice	1.94	2.30	4.51	3.87	4.17	5.64
Beans	.50	.52	.61	1.27	1.31	1.50
Corn	1.50	1.87	2.40	2.93	3.16	3.98
Wheat	1.26	1.41	2.20	4.99	4.99	4.99
Cartamo	.66	1.01	1.17	1.18	1.34	2.20
Sorghum	2.99	3.17	4.00	4.59	4.66	4.95
Sugar	66.73	67.40	70.14	81.53	84.40	96.91
Cotton	.45	.45	.45	1.08	1.08	1.09

Sources:
(i) 1985 yields/ha for irrigation were derived by applying the percent of each crop's 1981 production accounted for by irrigation *(See Table 4)* to 1985 production levels reported in *Programa. . . .*, p. 103, and dividing by acreage given in Part A. The remaining production is divided by the non-irrigated acreage given in Part A to determine 1985 yield/ha for non-irrigated farms.
(ii) Irrigation yields/ha for the year 2000 are estimated by Aguilera G., 1985, *op cit.*, p. 120. The calculated compound annual rate of increase in irrigated yield/ha between 1985 and 2000 is applied to estimate the 1988 level for yield/ha. To estimate 1988 yield/ha for the non-irrigated sector, total 1988 production for each crop, *Programa. . . .*, p. 103, is reduced by 1988 production from the irrigated sector (1988 yield/ha times hectares given in Part A) and divided by non-irrigated hectares for the crop given in Part A.
(iii) Yield/ha for each crop in the non-irrigated sector for the year 2000 is estimated by applying the 1985–88 compound annual rate of change to the 1988–2000 period.

As Table 10 indicates, most priority irrigation projects planned during 1986 (the last year for which complete data are available) involve *real* costs per hectare, which are at least 50 percent higher than comparable costs for developing irrigation lands during the 1960s and 1970s. *(See Table 8.)* These data suggest that Mexico's strategy for increasing agricultural production by expanding irrigated land may require careful rethinking. Annualized capital costs for new irrigated lands as shown above range from 231,000 to 984,000 pesos per hectare (costs in Table 10 annualized at 10 percent, a rate *well* below current rates now in excess of 45 percent). Annual average O&M costs are on the order of 58,000 pesos (1986 pesos) per hectare (Table 20). Thus, costs for adding new irrigated lands seem to range from 289,000 to 1,042,000 pesos per hectare. As Table 11 shows, the only crops in the Northwest that yield returns within this range of costs are beans (1.2 million pesos/hectare), sugar cane (1.2 million pesos/hectare), and cotton (2.7 million pesos/hectare), and these crops constitute a relatively small proportion of crops typically grown in the region. Indeed, beans account for only 59,000 hectares (3 percent of land in irrigated crops); sugar cane, 39,000 hectares (2 percent of irrigated acreage); and cotton, 235,000 hectares (12 percent of irrigated acreage).

Table 7. The Potential for Expansion in Mexico's Agricultural Land Base

Regions	Developed Agricultural Lands—1977 (1981): Irrigated	Dry Land	Total Arable Land: Irrigable	Non-Irrigable	Limits on the Expansion of Agricultural Lands in Mexico: Irrigable	Dryland
	(000 ha.s)		(000 ha.s)		(000 ha.s)	
Total Mexico	5,363 (5,497)	15,481 (17,626)	10,057 (53%)*	22,316 (69%)*	4,694	6,835
North	3,313	5,162	4,656 (71%)*	9,015 (57%)*	1,343 (29%)	3,853 (56%)
Central	1,608	6,007	2,201 (73%)*	6,628 (91%)*	593 (13%)	621 (9%)
South	412	3,551	2,799 (15%)*	4,110 (86%)*	2,387 (51%)	599 (8%)
Southern Peninsula	30	761	401 (7%)*	2,563 (30%)*	371 (8%)	1,802 (26%)

Source: Aguilera G., 1985, Table 13.

*Percent of total that was in production (developed) in 1980.

Similar crop mixes exist in the other regions. In the north-central region, only beans (48,000 hectares; 8 percent of irrigated acreage) and cotton (110,000 hectares; 19 percent of irrigated land) yield returns that might cover the costs of irrigating new lands. *(See Table 11.)* In the central and east regions, the only crop with such potential is beans, constituting but 3 percent of the irrigated lands in both regions. Land scarcity aside, any feasible program for expanding land under irrigation would require close consideration of how Mexico is to insure that such high-value crops as beans, sugar cane, and cotton are grown mainly on newly irrigated land.

Although the National Program implicitly calls for agricultural land expansion, land scarcity and high land-development costs argue for the second strategy specified in the National Program—increasing agriculture's efficiency and productivity. Certainly, relying on productivity increases would be consistent with past Mexican experience. As Table 12 shows, more than half of Mexico's output gains for corn and wheat between 1970 and 1985 reflect productivity gains. And the primary source of such gains in this period for *all* crops was irrigated agriculture. *(See Table 5.)*

In the drive to make resource use in irrigated agriculture more efficient, several caveats need to be heeded, three of which have already been discussed. First, the social and economic equity issues raised by investing in the most productive subsector of agriculture (irrigated) instead of in the largest and neediest (non-irrigated) cannot be ignored. Second, capital-intensive options like those employed in the short-lived Mexican Food System (SAM)

Table 8. Historical Investment Costs for Mexican Irrigated Acreage

Period	Area Benefited (000 ha.)	Total Investment (billions of 1986 pesos)	Cost/ha. (thousand 1986 pesos)
A. Small Scale Irrigation Projects			
1947–52	143.1	48	338
1953–58	155.9	43	273
1959–64	106.1	50	468
1965–70	84.7	143	1,683
1971–76	240.2	508	2,116
1977–81	321.6	738	2,294
Total	1,051.6	1,529	
B. Large Scale Irrigation Projects			
1947–52	377.2	657	1,742
1953–58	381.5	325	851
1959–64	105.6	275	2,601
1965–70	266.9	282	1,055
1971–76	618.5	590	955
1977–81	530.0	1,429	2,696
Total	2,277.7	4,338	
C. Total Irrigation Development			
1947–52	520.3	707	1,359
1953–58	537.4	367	683
1959–64	209.7	324	1,547
1965–70	351.6	424	1,206
1971–76	858.7	1,099	1,279
1977–81	851.6	2,164	2,541
Total	3,329.3	5,086	

Source: World Bank, 1983. 1981 pesos adjusted to 1986 pesos, Table 27.

should receive the lowest priority since financial "crises" are the order of the day in Mexico. And, third, these crises cannot be used as an excuse to ignore the irrigation-management problems that could undermine attempts to meet them.

These physical, social, and financial constraints, along with the land-tenure issues discussed in the next chapter, are clearly serious. Nonetheless, if increased agricultural production is the overriding goal, investments in improving the productivity of irrigated agriculture do represent the best way to achieve it.

Table 9. Marginal Costs of New Irrigated Lands in Mexico; Three-Year Moving Averages for 1979–82

Year	**Costs of Bringing New Lands into Irrigation Three-Year Moving Average**
	(thousands of 1986 pesos per hectare)
1979	1,688
1980	2,312
1981	2,503
1982	3,098

Source: World Bank, 1983, p. 105. 1981 pesos adjusted to 1986 pesos, Table 27.

Table 10. Current Proposals for Irrigation Projects

District	**Initial Investment**	**Area Irrigated**	**Cost Per Hectare**
	(billions of 1986 pesos)	**(ha)**	**(thousands of 1986 pesos)**
1. Nexapa (Sinaloa)	115	14,983	7,656
2. Elota Pioixtla (Sinaloa)	245	64,954	3,750
3. Ocoroni (Sinaloa)	81	11,235	7,188
4. Cuxtepeques (Tamoulipas)	52	10,410	5,000
5. San Gregorio (Tamoulipas)	44	4,400	9,843
6. Cuahuayana (Michoacan)	52	25,570	2,031
7. Autlan El Grullo (Jalisco)	46	18,284	2,531
8. Bajo Rio Bravo (Tamoulipas)	304	140,000	2,188

Source: Unpublished data obtained from SARH, September 15, 1985. 1981 pesos adjusted to 1986, Table 27.

Table 11. Net Returns to Land and Water for Selected Crops by Region, 1983. Values in thousands of 1986 Pesos

		NORTHWEST:				NORTH CENTRAL:		
Crop		**Net Returns Per ha.**	**Net Water Use/ha.**	**Net Returns Per cm.**		**Net Returns Per ha.**	**Net Water Use/ha.**	**Net Returns Per cm.**
			(cm/ha.)				**(cm/ha.)**	
Maize	(2)	50	34	1	(16)	39	41	1
Beans	(3)	1,200	35	34	(8)	1,072	37	29
Wheat	(27)	466	55	8	(10)	300	81	4
Sorghum	(3)	292	64	5	(5)	181	60	3
Sugar cane	(2)	1,155	131	9	(0)	—	—	—
Cotton	(12)	2,722	89	31	(19)	1,830	85	22

		CENTRAL:				EAST:		
Crop		**Net Returns Per ha.**	**Net Water Use/ha.**	**Net Returns Per cm.**		**Net Returns Per ha.**	**Net Water Use/ha.**	**Net Returns Per cm.**
			(cm/ha.)				**(cm/ha.)**	
Maize	(32)	115	38	3	(32)	23	55	.4
Beans	(5)	1,198	19	63	(3)	941	53	17
Wheat	(9)	292	66	4	(1)	146	58	3
Sorghum	(10)	270	31	9	(30)	140	38	4
Sugar cane		—	—	—	(7)	683	101	7
Cotton		—	—	—	—	—	—	—

Source: World Bank, 1983, pp. 51, 52, and Table 33(a). Numbers in parentheses indicate the percent that acreage in that crop is of total harvested acreage. 1981 pesos adjusted to 1986 pesos, Table 27.

Table 12. Estimated 1970–85 Production Changes Attributable to Acreage Expansion and Productivity Increases

Crop	Expanded Acreage	Productivity Increases
Rice	71%	29%
Beans	61%	44%
Corn	3%	97%
Wheat	43%	53%
Sorghum	80%	20%

Source: Based on data in Table 1, the following are computed:

$$\text{Inc (Prod)} = \text{Prod (85)} - \text{Prod (70)}$$
$$\text{Inc (Prod)} = A(85)y(85) - A(70)y(70)$$
$$\text{Inc (Prod)} = [\{A(85) - A(70)\}y(85) + \{y(85) - y(70)\}A(70)]$$
$$\text{PERCENTACE INC (PROD)} = \frac{\{A(85) - A(70)\}y(85)}{\text{Prod (70)}} + \frac{\{y(85) - y(70)\}A(70)}{\text{Prod (70)}}$$

where Prod(x) denotes production in year x; A(x) and y(x) denote acreage and average yield in year x. The two terms on the right hand side of the final equality give acreage and yield effects respectively. The authors gratefully acknowledge this expository suggestion by K. William Easter, University of Minnesota.

AN IMPORTANT CAVEAT: LOOKING BEYOND THE CURRENT FINANCIAL "CRISIS" IN MEXICO

During 1985, Mexico experienced a number of economic and natural crises, including the earthquake that devastated Mexico City and 64-percent inflation. In addition, the peso was devalued by several hundred percentage points from 1982 to 1985 (*Wall Street Journal,* February 3, 1986, p. 3). By 1986, the federal debt was approaching $101 billion, and Mexico had to borrow substantially larger sums, in excess of $5 billion, than were originally estimated (*Wall Street Journal,* February 5, 1986). Responding to these crises, the Mexican government imposed stringent capital controls. As a result, much of the private capital that Mexico needs has fled the country.

Clearly, the economic uncertainty[4] that continues to pervade Mexico makes policy options that involve, for example, capital infusions or increasing factor costs seem insensitive at best and unrealistic at worst. Yet, the country's growing needs for food and fiber are real and warrant Mexico's current concern—amid the immediate economic crisis—for at least two reasons. First, the awareness of such problems may alert policy-makers to important long-run implications of policy actions designed to deal with shorter-run problems. Second, even if the funds required to implement new policies are currently limited, planning for the means to deal with long-run problems in Mexico's irrigated agricultural sector should begin now.

III.

LAND TENURE—AN ABIDING CONCERN

Mexican agriculture has been shaped by the country's physiographic setting and its people's settlement patterns. In addition, since the Revolution of 1910, land reform and increased government involvement in food production have given a greater proportion of the population a stake in the agricultural economy. But—as a little history shows—some inequities in land distribution have persisted and continue to influence differences in income and in agricultural output between the private and *ejidal* landholders, which in turn influence the productivity of irrigated agriculture.

The growth of irrigated agriculture in Mexico has occurred in conjunction with the development of the country's land tenure system. When Spanish conquerors arrived in Mexico in the early 1500s, they found the best agricultural lands already settled by native tribes practicing fairly sophisticated cultivation techniques. During pre-conquest times, the lands and their produce typically belonged to the villages under a communal arrangement. However, during the process of "civilizing" and "Christianizing" the Indian populations, the Spaniards replaced the communal structures with their own *encomienda-hacienda* system—a rigid class structure in which predominantly Indian peasants labored under landowners of primarily Spanish descent. By the early 1900s, fully 90 percent of the rural population owned no land at all, and by the beginning of the Revolution in 1910, farmland was almost entirely in the hands of a few thousand landowners, some of whom owned several million acres each.[5]

Although an estimated 2.5 million acres of land was under cultivation at the time of the revolution, most of this had been developed by foreign land companies to grow such export crops as sugar cane and cotton. Aside from this use, most *haciendados* lands in the northern regions were left idle or cultivated unproductively.

Before the revolution, thousands of landless peasants sought the scarce, low-paying jobs on the estates. Rural poverty ran high, and landlessness became identified with economic insecurity. Not surprisingly, the slogan "the land belongs to those who work it" gained popularity during the political upheaval.

Agricultural reform began slowly after the revolution largely because political and social events overshadowed agricultural development. During the armed uprisings of 1910 to 1921, agricultural production fluctuated widely and output actually fell to levels below those of 1900 and 1902. By 1934, the revolutionary government was firmly in power and began rebuilding the nation's agricultural structure in accordance with the 1917 constitution. Major emphasis was given to land reform by the Cardenas administration (1934–40), which distributed 20 million hectares of land (65 percent of the total land distributed from 1917 to 1940) among 750,000 peasants.

Realizing that land reform alone could not greatly increase agricultural output, the Mexican government boosted agricultural credit during the late 1930s by initiating investment programs for irrigation development. The Cardenas administration founded the *Banco Nacional de Credito Ejidal* to promote agricultural activity among the newly emerging *ejidos.* During this period, 144,000 hectares of land was irrigated and water distribution was improved on another 103,000 hectares under government programs that marked the beginning of irrigation development. In subsequent years, however, this emphasis on land reform through irrigation projects was not matched by the necessary public investment.

The major thrust of the Constitution of 1917 was to initiate land redistribution by expropriating portions of large estates (subject to compensation) and distributing these lands to landless peasants who met certain criteria. Under this constitution, the government owned all unclaimed land and had the power to sell such land to private parties. Land could be condemned (with compensation) if required for public welfare, and the state was to regulate private land, when required, for the public interest. The state was also allowed to require the divestiture of large landholdings to create small agricultural population centers. This provision formed the basis for the revolution's basic program for agrarian reform—the right of all rural population centers to have lands, or *ejidos,* along with the protection of "small" private farms.

Given this history, what is the status of land reform today? First, even though a great deal of land was redistributed under the agrarian reform, inequalities in landholdings remain. Maldistribution is most evident in the irrigation districts, where a little over 100,000 private farmers cultivate nearly 3.7

EJIDOS

The *ejidal* system of land tenure refers to a village and surrounding lands where property is held in common. The establishment of an *ejido* is accomplished by an application submitted jointly by twenty or more villagers. Upon approval by the government, the village is granted a piece of land *(ejido)* from which individual members receive small plots. Within each *ejido,* an Assembly is organized to administer and monitor affairs of common interest to *ejido* members. Once the land grant is received, the Assembly determines the area of potential cropland and divides it into plots or parcels for distribution to each member *(ejidatario).* The *ejidatarios* do not pay for the granted lands. While they may pass their parcels on to their heirs, they cannot sell or rent their land. If they do not work the land themselves, they must forfeit it.

Originally, the size of *ejidal* parcels was set in the Constitution at four hectares of irrigated land or eight hectares of "seasonal" land (currently updated to ten hectares and twenty hectares, respectively). However, in the face of the population growth, the number of peasants entitled to receive land has been so large that the units distributed have actually averaged less than two hectares per family. Many peasants still have not received land and have been granted the status of "*ejido* members with reserved agrarian rights," which allows them to receive land in areas other than their village of origin.

The 1970 census recorded approximately 2 million *ejidatarios* residing in over 21,000 *ejidos,* and both figures have increased slightly since then. This implies an average *ejido* population of somewhat fewer than 100 members.

million hectares of irrigated land, implying an average farm size (37 hectares) almost four times greater than that of *ejidatarios* in the same districts.

Second, the reorganization of farm lands during land reform was based on the agrarian code of 1934. Private farmers were permitted to keep 150 hectares of farm land and to *choose* the location of this land—even if it meant retaining title to widely dispersed, non-contiguous parcels. The remaining lands were then distributed among the *ejidatarios.* In many cases, this redistribution system created a "crazy quilt" pattern of ownership that still persists: *ejidal* lands and private lands are intermixed (Senior, 1974, p. 91).

Third, land reform has encouraged overpopulation in some areas. Consider the Comarca Lagunera region in Northern Durango and Southern Coahuila, for example. When land reform was first instituted, lands were to be given to peasants who lived within seven kilometers of the land and had worked in the region for at least six months. But the census data included many migrant workers, and the lands were allocated to many more people than originally intended.

Fourth, in areas such as the Comarca Lagunera, private farmers can supplement allocated river water with groundwater to a much greater extent than *ejidal* farmers can. In 1983, for example, three-fourths of the irrigated acreage of the private farmers used groundwater, compared to one third of the *ejidal* acreage. Over the years, private farmers' access to groundwater has increased (Senior, 1974). In 1948, private farmers owned 62 percent of the wells and 30 percent of the land, while *ejidos* had 30 percent of the wells and 70 percent of the land. In 1974, some 70 percent of the region's 2,800 ground water pumps were in private farmers' hands (Senior, 1974 and Palacios Velez, 1986). This inequity stems in part from the original land-allocation process: private farmers removed pumps and other equipment from their land before giving it up.

Ultimately, land reform as carried out in Mexico has resulted in sharp disparities between *ejidal* and private farms in terms of their access to resources. In northern Mexico, private irrigated farms range from 20 to 60 hectares while *ejidal* farms range from 8 to 16 hectares (Yates, 1981, p. 78). Per hectare, the output of private farms is more than 10 percent greater than that of *ejidal* farms—either because private farmers often cultivate higher value crops (Yates, 1981) or because such critical resources as water and credit are inequitably allocated and the *ejidal* farmers lack technical knowledge and expertise (de Alcantara, 1976).

"Land reform as carried out in Mexico has resulted in sharp disparities between ejidal *and private farms in terms of their access to resources."*

Obviously, these four factors will influence the future productivity of irrigated agriculture. Indeed, the fragmentation of landholdings, the availability of an adequate supply of water, and inequities in the allocation of inputs, credit, and extension information are issues that must be addressed and reconciled if the intended goals and benefits of the government programs are to materialize.

The Administrative Structure of Mexican Agriculture

The development of irrigated agriculture in Mexico began nearly seventy years ago. In 1921, an irrigation directorate was established within the Department of Agriculture and Development to administer water resources. In 1926, the National Irrigation Commission was created to regulate irrigation, and its vigorous initiatives, beginning in 1945, led to rapid and sustained growth of output from irrigation. Since the mid-1940s, land reclamation through irrigation has been the Mexican government's single most important policy tool for promoting and expanding agricultural output.

Land and labor productivity rose as the use of capital inputs—including equipment, fertilizer, and machinery—grew in the irrigation sector. Agricultural credit provided by the government increased enormously, primarily because in 1947 President

Aleman created a cabinet-level Department of Water Resources (the *Secretaria de Recursos Hidraulicos,* SRH) to replace the National Irrigation Commission. Funds were allocated generously: annual irrigation investments in 1946 were three times the 1942 levels (double the 1942 levels in real terms).

In the early 1970s, the SRH was integrated into the Ministry of Agriculture to form Mexico's Ministry of Agriculture and Water Resources (*Secretaria de Agricultura y Recursos Hidraulicos,* SARH) and Mexico's use of long-range planning in water resources management began. These initial planning efforts culminated, in 1975, in Mexico's first National Water Plan (*Plan Nacional Hidraulico,* PNH; *see* SARH, 1981[A]). With a horizon that extends to 2000, this plan attempts to match estimates of future demands for water by residential, industrial, and agricultural entities with estimated future supplies and to specify the water-development projects needed to meet shortfalls. SARH also has primary responsibilities for managing Mexico's water resources.

SARH's management role depends upon which type of irrigation entity is involved. The facilities of *Irrigation Districts* (which include about 70 percent of Mexico's irrigated acreage) are constructed, managed, and maintained by SARH, which charges users fees. The administrative organization of Districts was designed to manage irrigation in the large *ejidos* common to northern and northwestern Mexico. In Sonora, for example, five Districts contain between 45,000 and 540,000 acres each, and in Sinaloa the average District size is 315,000 acres. SARH also administers *Irrigation Units,* which include about 30 percent of the approximately 5 million SARH-administered hectares under irrigation. Irrigation Units range in size from a few hundred to a few thousand acres, and most are located in the central states. In Irrigation Units, water-users operate, maintain, administer, and *allocate* water within the guidelines of an approved plan while SARH's role is limited to technical assistance, construction, and the upgrading of distribution systems.

IV.

IMPROVING IRRIGATION MANAGEMENT

Although significant productivity gains on irrigated lands are technically achievable, numerous ill-conceived or ineffective resource management practices and policies are taking a heavy toll on Mexico's agricultural productivity and reducing the availability of irrigation water. The role and performance of Mexico's Ministry of Agriculture and Water Resources (SARH), as well as of various practices and policies that play out at the farm, irrigation district, and national levels must be thoroughly re-examined if irrigated agriculture is to live up to its potential in meeting Mexico's growing needs for food and fiber.

"Often, introducing irrigation can make double-cropping possible."

On the Farm

At the farm level, the major resource management issues are cropping patterns, input mixes, resource use, and the farmers' technical capabilities and education levels.

Cropping Patterns. One promising way to enhance farm-level agricultural productivity is to boost crop intensities (through plantings) and to better match crop patterns to available resources. Often, introducing irrigation can make double-cropping possible. However, in all of Mexico's irrigated agricultural sector, average cropping intensity between 1970 and 1983 was only 120 percent, which implies that only 20 percent of Mexico's irrigated lands are double-cropped. *(See Table 13.)* In Mexico's arid northwestern and north central regions, where agriculture production depends totally upon the availability of water for irrigation and where water supplies are limited, one might expect lower intensities, perhaps just a single planting each year. In fact, however, the Northwest's crop intensities of 130 percent are above that average, while—at 105 percent—crop intensities in the North Central region are below the national average. In such arid areas, further increases in crop intensities may be obtained only by using water more efficiently on-farm and off.

In some of these two regions' Irrigation Districts, crop intensities have been increased by augmenting

Table 13. Crop Intensities in Mexico's Irrigation Districts by Area, 1981

Area	Districts	Crop Intensity*
Northwestern	13	130%
North Central	12	105
Central	32	112
East	17	106
Mexico, Total	74	120

Source: World Bank, 1983, p. 99. As used in this report, the Northwestern area includes the states of north and south Baja, California, Sonora, Sinaloa and Nayari. The North Central area includes the states of Chihuahua, Coahuila, Nuevo Leon, Tamaulipas, Zacatecas, and Durango. The Central area includes the states of Jalisco, Aguascalientes, Guanajuato, Michoacan, Mexico, Colima, and Guerrero. The East area includes the states of San Luis Potosi, Veracruz, Queretaro, Puebla, Tlaxicala, Oaxaca and the peninsula states.

*Hectares planted divided by irrigable hectares in the District.

scarce river waters with groundwater. In the Comarca Lagunera Irrigation District in northern Durango-southern Coahuila, some 26,000 hectares of the District's 104,000 irrigated hectares (averages from 1970 to 1983) are double-cropped. Water for crops, primarily summer crops, comes mainly from the Nazas River which irrigates about 70,000 hectares; another 34,000 hectares of summer crops (primarily alfalfa) is irrigated with groundwater (Palacios Velez, 1986). Land that is double-cropped (usually for winter wheat and alfalfa) is irrigated with groundwater.

Irrigation Districts in eastern Mexico enjoy relatively abundant rainfall—certainly enough for one crop during the wet season. Moreover, irrigation supplies permit irrigation during the dry season. Yet, as Table 13 shows, cropping intensities in the East are among Mexico's lowest. Although the reasons for this anomaly are not perfectly clear, a World Bank survey conducted in 1982 sheds some light (World Bank, 1983, p. 2). According to the report, in many areas (the Rio Panuco Irrigation District, for instance) farmers simply prefer to grow one crop during the wet season and (the availability of irrigation waters notwithstanding) not to farm during the dry season using irrigation. Apparently, they want to avoid plant diseases, the risks of flooding, and the hardships of working during rainy days. (World Bank, 1983, pp. 2–5).

Besides crop intensification, another means of increasing the productivity of land and water in irrigated agriculture is to optimize cropping patterns. Switching to crops that require less water can help increase productivity in water-short areas. In Mexico, farmers are not selecting crop patterns to maximize returns on land or water. *(See Table 11.)* If they did, beans and cotton would be major crops in Mexico's north central region. Right now, only 3 to 8 percent of the irrigated land in this area is in beans and only 12 to 19 percent in cotton. Moreover, 16 percent of the north central region's irrigated land is planted in maize that yields only 1,000 pesos per centimeter of water—20 times less than beans or cotton do.

"Productivity gains may well attend Mexico's efforts to choose crop mixes that better reflect the relative conditions of scarcity of land and water resources."

Table 11 also substantiates the often-heard recommendation that production of such water-intensive crops as sugar cane should be shifted to the relatively water-rich and comparatively humid eastern region where returns on water use would be relatively high. *(See,* e.g., Barkin, 1983.) Thus, productivity gains may well attend Mexico's efforts to choose crop mixes that better reflect the relative conditions of scarcity of land and water resources.

Input Mixes. Mexico was effectively the cradle of the Green Revolution (Barkin, 1983). With help from the Rockefeller Foundation, Mexico created an Office of Special Studies as a part of its State Department in the early 1940s to promote joint American-Mexican research on improved seeds and hybrid varieties of various crops. With wheat, Mexico's program was notably successful: average yields rose from some 750 kilograms per hectare in the early 1940s to almost *4 tons* per hectare in 1980. This "success" may have occurred largely because wheat is grown almost exclusively under irrigation, primarily by comparatively well-educated farmers who are far more likely than their counterparts in rainfed areas to use fertilizers and the other inputs needed to get high yields from hybrid seeds.

Unfortunately, parallel yield increases have not been obtained from improved seeds for other crops. Between 1940 and 1980, corn yields rose from 600 kilograms/hectare to less than 1,800 kilograms/hectare, and bean yields doubled from 225 to 550 kilograms/hectare (Barkin, 1983, p. 14). Gains in yields have been relatively slow because such requisite inputs as fertilizers have been unavailable throughout Mexico's temporal agricultural areas—where the bulk of these crops is grown—and because, as a result, few Mexican *campesinos* have been able to adopt hybrid corn and bean seeds.

With or without hybrid seeds, Mexican farmers in the irrigated and rainfed sectors, particularly those far from urban centers, have no dependable supplies of "modern" agriculture inputs. As the World Bank reports:

> . . . *many farmers are without a regular supply of seeds, fertilizers, machinery, plant protection, credits, and crop insurance. Although Mexico's efforts since the mid-1970s to correct this have been impressive, entities differ in the quality of services because of lack of communication, experienced staff, storage facilities and equipment.* (World Bank, 1983, p. 4.)

Farmers lack adequate access to credit too. Mexico's long commitment to making credit available to agriculture dates back to 1935, when the National Agricultural Credit Bank was established. During the 1940s and 1950s, numerous agricultural credit institutions were created, including the National Union of Sugar Producers, the National Foreign Commerce Bank, and the National Agriculture Insurance Agency. Since the 1970s, the Mexican government has attempted to centralize agricultural credit. In 1975, most banking institutions issuing agricultural credit were merged into the National Rural Credit Bank.

Many forms of agricultural credit have long been subsidized in Mexico, though this subsidy has eroded over the last decade as the country's general financial difficulties have deepened. Subsidies through credit grew, in nominal terms, through the 1970s, falling only in 1981. In *real* terms, however, credit subsidies peaked in 1979 and have been falling since. *(Table 14.)*

Despite the decline in real credit subsidies, Mexico's central bank has been remarkably successful in keeping credit available to the agricultural sector. As Table 14 shows, despite a few tight years, the central bank managed to maintain—and increase somewhat—the real amount of credit available to agriculture (per harvested hectare) through the 1970's and early 1980's. Still, interviews with personnel with the central bank during 1985 suggest that the availability of real credit in agriculture has decreased dramatically since 1982—not surprising, given the expansionary, inflationary, and monetary effects of maintaining real credit to this sector.

Education, Extension, and Research. Education levels are generally low in Mexico's agricultural sector. High school completion rates in Mexico's rural areas range from 20 percent to 32 percent, compared with the national average of 51 percent (SARH, PNDRI, June 1985, p. 31). Some 60 percent of Mexico's rural population cannot read or write (SARH, June 1985, p. 31)—one reason that technological advance in the agricultural sector is slow.

For some time, the Mexican government has sponsored extension and research programs designed to upgrade farmers' technological capabilities. Since 1981, the government has spent some 2.9 billion pesos (1986 pesos) per year on agricultural extension. Typical of successful new extension

Table 14. Total Credit Made Available Through Mexico's Central Bank for the Agricultural Sector, 1970–82

Year	Agricultural Credit[1] Made Available Through Mexico's Central Bank	Agriculture Credit[2] per Harvested Hectare
	(billions, 1986 pesos)	(thousands, 1986 pesos)
1970	1,790	120
1971	2,038	132
1972	2,105	138
1973	2,114	133
1974	2,248	151
1975	2,210	143
1976	2,200	149
1977	2,371	142
1978	2,676	162
1979	2,905	195
1980	1,933	108
1981	3,571	191
1982	3,457	182

1. Agricultural credit as a percent of GNP was obtained from Ruiz Duran Clemente "La Politica Monetaria y Crediticia 1970–81," pp. 75, 77. GNP was taken from Instituto Nacional de Estadistica, Geografia E Informatica, *Sistema de Cuentas Nacionales de Mexico,* Mexico, D.F., 1983, p. 1. 1970 pesos adjusted to 1986 pesos, Table 27.
2. Harvested hectares taken from World Bank, 1983, p. 84.

efforts is the Mexicali Irrigation District in Baja California Norte. *(See Table 15.)*

For several reasons, however, most expenditures on extension activities have not been notably productive. First, each extension officer must serve all the farmers on 2,500 hectares of land and a large proportion of the extension officers have no training or experience. Second, much of the agricultural research underlying extension services has limited relevance for most Mexican farmers. During the early 1980s, for instance, Mexico's National Center for Advanced Methods of Irrigation (CENAMAR) in Torreon, Coahuila, spent more than half of its limited research budget on irrigation systems other than gravity irrigation, even though more than 95 percent of Mexico's irrigation is gravity-fed. On-farm storage of water, the development of crops that need little water, and tilling and seeding

Table 15. Extension Agents and Expenditures for Extension in the Valle de Mexicali, 1981–86

Year	Number of Extension Agents	Expenditures for Extension Work
		(millions of 1986 pesos)
1981	n.a.	39,700
1982	96	57,100
1983	104	96,500
1984	n.a.	n.a.
1985	126	272,500
1986	137	398,000

Source: Trava, 1986.

practices—all of which can increase net farm income and make resource use more efficient have also been neglected (Annex 1 of World Bank, 1983, pp. 4, 5 and 22).

Two examples of the kind of "strategic" research needed now illustrate its use and importance. The first concerns the use of plastics for mulch. Yield increases in field experiments with plastic mulches typically range between 10 and 25 percent; under controlled experimental conditions, increases of 50 to 100 percent have been realized (Anderson and Fernandez, 1983; Nicholas, 1983). Such yield increases have been attributed to reduced leaching of nitrates (Clarkson, 1960), moisture conservation (Harris, 1965), and the elevation of carbon-dioxide levels in the microclimate. (Shelddrake, 1963). Besides increasing crop levels for the individual farmer, the use of plastic mulch can also shorten the crop cycle. In France, for example, when plastic mulch was used on maize, seed germination decreased from 20 to 15 days, and the harvest took place three weeks earlier than normal (Spice, 1977). In controlled experiments in South Africa, carrots grown under plastic mulch matured two weeks earlier, radishes one week, lettuce and cabbage two weeks, and rhubarb up to four weeks earlier than normal (Nicholas, 1983). At the Espanola Valley Branch Experiment Station in New Mexico, plastic mulch treatments speeded up the early growth of tomatoes, cantaloupe, and watermelon and thus raised early yields (Trujillo and Corgan, 1981). Also, Hopen and Oebker (1976) cite over 200 research studies on the use of plastic mulches in vegetable crop production; and most of these studies confirm earlier maturity and greater growth in response to mulching.

Clearly, Mexico should consider expanding research now under way at its *Centro de Investigacion en Qimica Applicada* on plastic mulch. Balance-of-payments problems might ease if plastics were used and revenues from products exported to the United States rose accordingly. Mexico also needs to find petrochemical applications that broaden oil production's contribution to the rest of the economy. More important here is Mexico's concern with how to increase agricultural production in response to its growing needs for food and fiber. One recent study (Cummings and Brajer, 1989) of how the timing of Mexico's irrigated tomato harvest in Sinaloa and Sonora influenced the prices received for tomatoes in the U.S. market indicated that using plastic mulches could increase Mexican revenues by some 10 percent—from $227.65 million to $250.67 million. More generally, growing evidence suggests that, for many of the crops that earn scarce foreign exchange for Mexico, strategic planning for the *timing* of exports—and the attendant implications for production technologies—may have high payoffs.

A second example of how practical strategic research could pay off in Mexico is of "water harvesting"—the process whereby precipitation runoff is increased and facilities are constructed to collect and store it (*see* U.S. Department of Agriculture, 1983). So far, water-harvesting techniques have been used mainly in Mexico to provide drinking water and water for domestic animals, but efforts are needed to expand the use of water-harvesting technology—especially that for reusing tail waters—in irrigation.

Of course, such research per se, however "strategic" or timely, cannot make a difference unless it reaches the Mexican farmer. As noted earlier, while real expenditures for education and extension programs in Mexican agriculture have been reasonably maintained over recent years, the quality of most programs (with notable exceptions in the Baja California Norte area) has deteriorated.

In the Irrigation District and Unit

Two issues of primary interest at the level of the irrigation district and unit are the maintenance and rehabilitation of irrigation facilities and SARH's allocation of water resources.

In most Districts, large surface-storage facilities are operated in association with complex systems of off-farm main, secondary, and tertiary distribution canals, and large-scale drainage systems, roads, and other infrastructure. Most of the irrigation units that make up the remaining 30 percent of the irrigated lands are constructed by SARH, but maintained and managed by the units' members. Units are usually designed and planned *ad hoc* and locally: most rely on groundwater for irrigation; the rest draw from surface storage facilities, river-diversion dams, natural springs, and the like. Except in the North Central region, average farm size is typically smaller in units,— around 3.67 hectare/family, compared with 6.13 hectare/family in districts. *(See Table 16.)*

In general, surface-storage facilities constructed by SARH are very well designed, especially the large-scale facilities found in districts (World Bank, 1983, p. 7). But most irrigation projects in Mexico are in one way or another incomplete—sometimes by design and sometimes not. In many cases, SARH intentionally postpones, or leaves to local authorities or farmers, minor or less expensive parts of an irrigation project (the construction of roads or drainage facilities, for example). In others, unanticipated problems—soil alkalinity, the dwindling of water resources, and soil erosion—arise after a project is completed and makes further investments necessary. Obviously, it's difficult to improve field efficiencies, which are as low as 45 percent between farm gates and main canals in some districts (SARH, 1982, p. 33), if delivery canals need repair and if facilities for controlling and measuring water deliveries and other works remain unfinished. Efficiency is particularly undermined when, as often happens, the funds needed to complete the project are appropriated from maintenance funds, thus short-shrifting maintenance and inviting the overall project to deteriorate (SARH, 1982).

Table 16. Average Productivity Measures for Districts and Units in Mexico's Irrigated Agricultural Sector, 1981

Area	Average Farm Size	Water Use Per Hectare	Crop Value Per Hectare	Crop Value Per Cm
	(Ha.)	(cm)	(thousands of 1986 pesos)	(1986 pesos)
Mexico:	5.02	62.73	466	7,429
Districts	6.13	65.18	437	6,704
Units	3.67	58.02	541	9,324
Northwest:	11.82	67.53	423	6,264
Districts	13.03	67.83	419	6,177
Units	5.78	63.71	482	7,566
North Central:	5.05	71.94	561	7,798
Districts	4.64	79.74	617	7,738
Units	5.61	64.88	493	7,599
Central:	2.80	60.03	551	9,179
Districts	3.81	70.75	517	7,307
Units	2.79	51.38	591	11,503
East:	6.61	49.10	373	7,597
Districts	9.13	35.28	310	8,787
Units	4.53	63.80	502	7,868

Source: SARH, "Informes de Riego," DGDUR, Mexico, D.F., 1981. C. 1981 pesos adjusted to 1986 pesos, Table 27.

Activities designed to "complete" the irrigation project are referred to as a project's "rehabilitation." The scale of such activities in Mexican irrigation is staggering: between 1965 and 1981, more than one fourth of all irrigated lands and one third of the budget for SARH's Subsecretariat of Hydraulic Infrastructure, which handles new construction projects, were earmarked for rehabilitation (SARH, 1982). With the increasing scarcity of funds and the growing urgency of completing projects, much of the *de facto* responsibility for project rehabilitation has shifted to the General Directorate of Irrigation Districts and Units (DGDUR) within the Subsecretariat of Agriculture and Operation, which has primary responsibility for operating and maintaining (O&M) facilities. As a result, maintenance funds have been siphoned off into rehabilitation (which SARH describes as the "deferral" of maintenance), thus undermining efficiency. (SARH, May 1985, Chapters II and III; and SARH, 1982).

Which services are being deferred and at what cost? As Table 17 shows, annual maintenance activities deferred between 1977 and 1981 included work on the lining of some 600 kilometers of canals, and the repair of 1,300 kilometers of canal and 2,200 kilometers of roads. The accumulated cost of such deferred maintenance, according to the World Bank, is high:

> *"Although no precise data are available on the amounts and the volumes of work needed (for deferred maintenance), extrapolation from the estimates made by the DGDUR field staff from sampling of seven districts and eight units (covering some 18 percent of existing irrigation areas) indicates that overall cost of on-and-off-the-farm works could be in the order of $3.5 billion (1981 dollars), or an average of about $700/ha. for the total of close to 5 million irrigated ha in Mexico."* (World Bank, 1983, p. 8.)

As for rehabilitation, annual expenditures for a seven-year plan were estimated by World Bank personnel as 302 billion pesos (1986 pesos, or $0.12 billion at the 1989 average free rate of exchange 2,481 = $1). The total rehabilitation "debt" for seven years would thus total some 2,114 billion 1986 pesos. *(See Table 18.)* Most project completion needs are for conveyance networks—especially those in eastern Mexico, where rehabilitation expenditures on the order of 157 billion pesos per year are proposed for the seven-year planning period.

This rehabilitation debt can compound itself in decreased production losses due to high conveyance losses of water and to soil-quality deterioration. In the Culiacan irrigation district, for instance, salt and soda concentrations in irrigated lands rose substantially between 1970 and 1985, affecting about one fourth of the district's usable lands, because drainage facilities were incomplete. *(See Table 19.)* Average yields for one of the area's major crops—soybeans—declined from 2.8 tons/hectare in 1970 to 2.0 tons/hectare in the early 1980s (SARH, September 1983, p. 59; and SARH, 1981, C. p. 90).

Estimates for O&M expenditures in several Mexican irrigation districts, given below in Table 20,

Table 17. Maintenance Works in Mexican Districts and Units Which Were Deferred, 1977–81

Item	Unit	1977	1978	1979	1980	1981
Canal lining	km	603	477	625	548	617
Earth canal repair	km	447	792	511	1,102	1,305
Repair of drains	km	646	957	1,374	1,608	2,200
Road works	km	403	266	575	524	7,430

Source: SARH, 1982.

Table 18. Planned Annual Rehabilitation Expenditures for Irrigation Areas in Mexico by Region, 1982

Rehabilitation Expenditures for:	Planned Annual Expenditures In: Northwest	North Central	Central	East
	(millions of 1986 pesos)			
Storage/diversion dams	150	600	450	8,333
Conveyance Networks	70,440	8,290	19,695	157,140
Drains	2,050	—	1,410	8,450
Roads	800	950	4,690	3,310
Other	1,690	690	3,960	8,550

Source: World Bank, 1983, pp. 23–29. 1982 pesos adjusted to 1986 pesos, Table 27.

Table 19. Soil Quality Changes in Irrigation Areas, 1970–85

Irrigation Area/Soil Type	Area	Surface Area in: 1970	1985
	(Hectares)	(Hectares)	
Culiacan/Humaya	350,000		
Normal		301,350	266,700
Saline		18,660	46,900
Saline-sodic		27,540	32,200
Sodic		2,750	4,200
Rio Yaqui	220,296		
Normal		192,138	198,307
Saline		7,628	8,315
Saline-sodic		17,325	10,624
Sodic		3,205	3,050
Rio Colorado	207,500		
Normal		95,425	126,020
Saline		86,110	80,630
Saline-sodic		67,665	41,760
Sodic		800	1,590

Source: Trava, 1987.

vary from 19,594 pesos/hectare to 115,625 pesos/hectare (1986 pesos) among the nine districts, while average O&M expenditures are 57,797 pesos/hectare. As Table 20 implies, water users in irrigation districts pay anywhere from 6 to 40 percent (an average of 27 percent) of O&M costs.

While comparable data for irrigation units are not available, in comparison, the Las Tuzas unit in Jalisco, where average farm size is about 3.3 hectares and water is distributed by gravity, annual O&M costs are but 43,000 pesos/ha. (1986 pesos)—all paid by water users. In the nearby Las Lagunillas unit, where groundwater is used, O&M costs are correspondingly lower: around 12,600 pesos.

What do these differences imply about efficiency? Since users pay for most O&M costs in irrigation units, water use per hectare is generally lower in irrigation units than in districts. *(See Table 16.)* Also, except in the eastern and north central regions, crop value per hectare and crop value per unit of water used are typically higher in irrigation units than in districts.

On lands managed by SARH, increases of salt and soda in the soils have reduced yields in many irrigation areas, mainly in Culiacan and other irrigation districts, where drainage facilities are inadequate, but also in some areas where low-quality water is used for irrigation. However, between 1970 and 1985, in areas such as the Rio Yaqui and the Rio Colorado, soil quality improved as a result of SARH's land-rehabilitation programs—mainly, extensive drainage works. As Table 19 shows, such efforts helped increase the amount of "normal" lands in the Rio Colorado area by more than 25 percent between 1970 and 1985.

At the National Level—Groundwater Management and Water Pricing

At the national level, the two main impediments to improved irrigation are poor groundwater

Table 20. Expenditures for Operation & Maintenance in Mexico's Irrigation Districts

Districts	Total Costs	Area in Hectares	Cost per Hectare	Expenditures Paid By SARH	%	By User
	(mill. 1986 pesos)		(1986 pesos)	(per ha.)		(per ha.)
Rio Mayo	3,594	94,262	38,125	22,656	59	15,649
La Antigua	1,078	15,098	71,406	62,094	87	9,313
Morelia	3,125	45,998	67,938	64,188	94	3,750
Rio Fuerte	5,094	259,968	19,594	12,969	66	6,625
Lagunera	8,133	120,609	68,922	53,641	78	15,625
Culiacan	6,000	233,010	25,750	19,266	75	6,313
Don Martin	2,891	25,000	115,625	106,250	92	9,375
Soto La Marina	1,500	26,016	57,656	52,875	92	4,781
Rio Yaqui	12,406	225,057	55,125	33,672	61	21,453
Totals:	44,000	1,044,418				
Averages:			57,797	25,688	73*	9,656

Source: World Bank 1983, App. III, p. 10. 1981 pesos adjusted to 1986 pesos. Table 27. O&M costs as reported here include "recurrent" costs; see Table 25.

*Total costs, 44 billion pesos, divided into total costs borne by the SARH: 32 billion pesos.

management and water prices that send farmers the wrong signals.

"At the national level, the two main impediments to improved irrigation are poor groundwater management and water prices that send farmers the wrong signals."

SARH's Groundwater Policies. The reliance of Mexico's irrigated agriculture on groundwater has increased markedly over the past three decades. In the 1950s, less than 5 percent of irrigation water was obtained from groundwater sources. By the late 1970s, that share had risen to about 35 percent—some 15.8 billion cubic meters/year (SARH, April 1984, pp. 164–68). In response to this increase, a new division *(Direccion de Aguas Subterraneas)* was created within SARH in 1964 to develop information and management policies for Mexico's groundwater reserves.

Unfortunately, SARH's groundwater division has continually suffered from a lack of funding for staff, fieldwork, and the test drilling needed to monitor groundwater stocks and collect the data policymakers require. SARH monitors groundwater pumping levels in a few areas, but both the agency's data collection and groundwater-management strategies compare poorly with SARH's control and management system for surface waters.[6]

These institutional deficiencies must change soon given the rising extent and importance of groundwater "mining." Pumping groundwater out faster than an aquifer can recharge has already become a serious problem in the Comarca Lagunera region in Northern Durango and Southern Coahuila, the Costa de Hermosillo region in Western Sonora, the Valle de Santo Domingo region in Baja California, and in the Valles de Leon and de Celaya regions in Guanajuato. (Chavez G. and Ramirez L. in *Consejo Nacional de Ciencia y Tecnologia, 1986.*) As Table 21 shows, in the Comarca Lagunera district and the Costa de Hermosillo, annual pumping exceeds annual recharge by 137 and 122 percent, respectively. In the Santo Domingo district, pumping exceeds annual recharge by more than 50 percent.

Crop intensities in the Comarca Lagunera district could not have increased if groundwater had not been tapped, but these gains have come at a price. The water table is now falling by some 1.5 meters per year. As a result, costs are increasing rapidly and fear of aquifer contamination or exhaustion is growing. Over the last 20 years, more than 2,700 wells have been drilled, almost three fourths of them by private farmers, to irrigate some 60,000

Table 21. Groundwater Mining in Selected Irrigation Districts, 1983

Region	Annual Pumping	Annual Recharge	Mining
		(million cubic meters)	
Comarca Lagunera	1,189	500	689
Costa de Hermosillo	777	350*	427
Santo Domingo	274	150	124

*Escamilla and Pinon report annual recharge in the Costa at 498 million m³/year. Cummings (1974, p. 14) however, reports 350 million m³/year as measuring natural recharge *including* return flows from irrigation—an estimate which current discussions with SARH personnel suggest as more accurate for the Costa's recharge.

Source: Escamilla and Pinon, 1983.

hectares (Palacios V., 1985, p. 35). As Table 21 shows, some 1.2 billion cubic meters are pumped in this region, 60 percent of which count as mining. Tables 22 and 23 indicate that the area's water tables have fallen from depths of 75 meters in 1965 to 106 meters in 1984 and pumping costs have risen correspondingly. Real pumping costs have risen from about .03 pesos (1986 pesos) per cubic meter pumped in 1965 to 2.78 pesos per cubic meter pumped in 1984 (*Consejo Nacional*, 1985, pp. 43, 52). Further complicating matters, pockets of contaminated groundwater were recently found in the outskirts of Torreon, Coahuila. Little is known yet about the contaminated plume spread, much less its cause.[7]

Significantly, the rate of mining in the Costa de Hermosillo has fallen since the early 1970s. Between 1970 and 1972, groundwater was pumped at the annual rate of some 1.2 billion cubic meters, which exceeded recharge by some 850 million cubic meters per year, in contrast to the 427 million cubic meter/year mining rate shown for 1983 in Table 21 (Cummings, 1974, chapter 2).

To deal with overdrafts, in the 1970's the Mexican government considered a massive south-to-north interbasin transfer of some one billion cubic meters of water from the Rio Fuerte in Northern Sinaloa. After several studies demonstrated the infeasibility of this plan—the Northwest Water Plan—it was abandoned in the late 1970's (Plan Hidraulico del Noroeste, PHLINO; Cummings, 1974). In 1975, SARH's Hermosillo office took another approach—requiring users to pump less groundwater and imposing heavy fines equal to the estimated *gross* value of the water on water use in excess of assigned quotas. The goal in 1975 was to reduce pumping levels to 600 million cubic meters/year within ten years and to eliminate mining within 20 years. The 1975–80 period was marked by radical changes in crop patterns (fruits and vegetables have increasingly become the important high value crops in the region) and substantive investments in canal linings (interviews with the District Manager, SARH office, Hermosillo, Sonora). These impressive efforts, while not totally consonant with the 1975 goals, helped reduce mining from 850 million cubic meters in the early 1970s to 427 million cubic meters in 1983. Further reductions in mining have occurred over the last few years.

As successful as this policy has been in reducing the rate of groundwater mining in the Costa de Hermosillo, it has had undesirable economic, social, and environmental impacts—including land abandonment and soil erosion. *(See Appendix.)* Of course, the need to husband groundwater *per se* does not explain SARH's interest in controlling pumping rates in the Costa de Hermosillo. As detailed in Cummings (1974), seawater encroachment into coastal aquifers has forced SARH to control groundwater pumping.

Table 22. Average Depth to Water Tables in the Comarca Lagunera Region, 1965–84

Year	Depth to Water Table
	(meters)
1965	75
1970	80
1975	90
1980	100
1984	106

Source: Consejo Nacional, 1986, pp. 19, 43.

Table 23. Average Pumping Costs in the Comarca Lagunera, 1965–84 (1986 pesos)

Year	Average Pumping Costs
	(pesos/cubic meter)
1965	.03
1970	.08
1975	.14
1980	.56
1984	2.78

Source: Consejo Nacional, 1986, p. 52.

In some areas, *rising* water tables associated with poor drainage and excessive water-use present SARH with serious problems, including potentially hazardous phreatic buildups. As Table 24 indicates, the amount of land in the Culiacan/Humaya and Rio Colorado irrigation districts with water tables within 1.5 meters of the surface almost tripled between 1975 and 1986. Consequently, SARH is now trying to formulate groundwater use regulations for these areas (Trava, 1987).

SARH's lack of control over groundwater use in Comarca Lagunera contrasts sharply with the actions SARH has taken in Costa de Hermosillo. Over the last three years, however, SARH offices in the Comarca Lagunera district have been promulgating management policies for groundwater use in the region. Under study are means for artificially recharging the aquifer (Petyjohn, 1985), for increasing pump efficiency (Moore, 1985), and for improving water use management in the region (Merkley and Keller, 1985; and Flores Lui, 1985).

Water Pricing. Looking for ways to induce irrigation users to use water and other resources more efficiently, many analysts cite the efficiency effects of pricing water (or any other resource) at its "scarcity" value (Howe, 1980). (Resources so priced are used only when their yield values equal or exceed their scarcity values.) In Mexico, irrigation is highly subsidized, even though there are no private property rights to groundwater or surface water. According to the law, which is implemented mainly by SARH, water users are *required* to repay parts of all capital costs associated with water-reclamation projects and *all* O&M costs. In terms of capital costs, users must—after a three-year grace period—repay 30 percent of investment costs at 12-percent interest over 22 years. In theory, then, the state can impose whatever regulations or prices on water use that it chooses to whatever ends—for example, cost recovery or water management.

Table 24. Depth to Water Table in Selected Districts in Northern Mexico, 1975 and 1986

Irrigation Area	Percent of Area's Land with Depth to Water Table Less Than 1.5 Meters in: 1975	1986
Culiacan/Humaya	2.7%	9.3%
Rio Yaqui	10.1%	11.0%
Rio Colorado	6.3%	16.6%

Source: Trava, 1987.

Despite this mandate, SARH has made no observable effort to recover the capital costs it incurs for irrigation projects in Mexico's irrigation districts. On the other hand, water users in irrigation units pay an average of about 30 percent of initial capital investment costs (Brajer and Cummings, 1986), and water users in districts and units pay 27 percent and 100 percent, respectively, of O&M costs. As Table 25 indicates, on average, water users in Mexico's irrigated agricultural sector pay only some 11 percent of the full costs of water. Equally important, imposed water prices, however low, are seldom linked to the quantity of water used; instead, in effect, they are flat taxes.

Subsidies at this order of magnitude (about 89 percent) give rise to at least three interrelated sets of effects. First and most obviously, the water subsidy distorts the relative prices of inputs used in agricultural production, resulting in efficiency losses. Since water is all but free, Mexican farmers use too much of it and, in some instances, too much of the fertilizers and other inputs that complement it. These other inputs are also often subsidized, further distorting the farmers' perception of efficient input use. A corollary effect, noted above, is subsidies' influence on crop patterns: with water priced artificially low, farmers have no incentives to look into water-conserving crops.

Second, underpriced water can encourage farmers to neglect the maintenance and repair of on-farm water-distribution systems. As a result, distribution efficiencies are low and, in many cases, falling.

Third, as noted earlier, subsidies for expanding irrigation increase budgetary outlays for this sector.

Table 25. Summary of Subsidies in Mexico's Irrigated Agricultural Sector, thousands of 1986 pesos

Marginal Costs for Irrigated Acreage[1]		**User Payments for:**	
Capital Costs[2]	376	Capital Costs	40
O&M Costs[3]	19	O&M Costs	10
Recurrent Costs[4]	47	Recurrent Costs	0
Total Costs:	442/ha.	Total User Payments	50/ha.
		Percent of Total Costs	11%

Implied Average Subsidy: *89 percent of total annualized costs*

1. See Table 6 in Brajer and Cummings, 1986; lower bounds are used for the purposes of this example. Percent subsidies would be unaffected by the use of upper bound data, of course.
2. Of a "representative" hectare, on 35 percent of the hectare (the unit portion) users pay 30 percent; users pay nothing on the 65 percent, district portion. Thus, users pay [(.35)(.3) + (0)(.65)] [capital costs], or 40,000 pesos per hectare.
3. Units pay 100 percent; districts pay 27 percent (Table 16); thus, on average, users pay [(100%)(.35) + (27%)(.65)] [O&M Costs], (52.55%)(19,000), or 10,000 pesos per hectare.
4. Neither units nor districts appear to pay recurrent costs. Recurrent costs represent large scaled O&M or "rehabilitation" activities such as upgrading dams and tubewell systems.

Source: Brajer and Cummings, 1986, Table 11, adjusted to 1986 pesos, Table 27.

They also further skew the distribution of incomes between the irrigation and rainfed agricultural sectors, and between districts and units. According to the World Bank,

> "*. . . because the average farm size in districts is higher than in units, the average irrigation transfer per beneficiary in districts was 2.6 times that of beneficiaries in units (in 1982)*" (World Bank, 1983, p. 66).

Such subsidies also have obvious opportunity costs. The subsidies must be financed either through other public projects foregone or through inflationary expenditure.

Using water pricing to remove subsidies and thereby improve the efficiency of water management in Mexico's irrigated agricultural sector is certainly not a new policy option for SARH. Indeed, as a result of a 1981 study of water charges by SARH's various district offices, new rates were imposed in many districts to capture cost increases (thereby reducing implicit subsidy rates). In virtually all cases, however, the fee increases left water priced at rates well below either its scarcity value or its real supply price. For example, in the Morelia Irrigation District, water rates high enough to cover full O&M costs were proposed (that is, only capital costs were forgiven) but were rejected at the Directive Committee level. *(See Table 26.)* Instead, SARH settled upon new "intermediate" rates that were much lower than the proposed rates but three to four times higher than the previous charges. Thus, for example, with the full cost of producing

"If water-pricing schemes are to be used to foster greater efficiency in water use, Mexican agricultural price policies in general have to be re-examined."

Table 26. Increases in Irrigation Fees

District	(Pre–1981) Previous Rate		(1983) New Rate
	(pesos)		(pesos)
4 (Don Martin)	375/ha basic rate		1200/ha basic rate
25 (Baja Rio Bravo)	375/ha irr. by gravity		1200/ha irr. by gravity
	60/ha irr. pumped from Rio Bravo		200/ha pumped from Rio Bravo
34 (Estado de Zacatecos)	600/ha basic rate		2500/ha basic rate
	Previous rate	**Proposed rate**	**New rate**
	(pesos)	(pesos)	(pesos)
20 (Morelia)	65/ha maize	1,516	125
	65/ha wheat	2,986	200
	160/ha vegetables	7,534	400
	250/ha strawberries	10,960	650

Source: Brajer and Cummings, 1986, Table 10.

Table 27. Wholesale Price Indices for Mexico: Selected Years 1980 = 100

Year	Wholesale Price Index
1960	16.1
1970	20.3
1981	124.4
1982	194.2
1983	402.7
1984	686.0
1985	1,093.4
1986	1,932.4

Source: International Financial Statistics: 1987 Yearbook, International Monetary Fund, Washington, D.C., 1987, pp. 494–495.

maize in Morelia equal to $1,516 pesos/hectare, the effective subsidy has been reduced from 96 percent to approximately 92 percent by the rate increase, even though actual water charges have nearly doubled.

The case for pricing water at "efficiency rates" may be weakened, however, by Mexico's policy of providing low-cost food and fiber to consumers by regulating the agricultural prices that farmers receive. Higher-cost inputs in conjunction with low, fixed-output prices could place Mexican farmers in a vise, so if water-pricing schemes are to be used to foster greater efficiency in water use, Mexican agricultural price policies in general have to be re-examined.

V.

POLICY RECOMMENDATIONS

Relatively rapid population growth and growing concern for the dietary needs of the rural poor make it urgent for Mexico to increase its production of food and fibers. To increase agricultural production even modestly for the next few decades, Mexican policy-makers must make hard decisions about expanding the Mexican agricultural land base and enhancing productivity on existing farms—irrigated and rainfed. Undeveloped arable lands in Mexico are scarce and, even if totally developed, would not completely answer Mexico's growing need for agricultural output. Moreover, developing many of these lands, particularly those that are potentially irrigable, may be prohibitively costly.

On Mexican irrigated farms, especially on district farms, subsidies destroy incentives for water conservation and water is treated as essentially a "free" good. As a result, water is wasted in overly large field applications, unmaintained canals, and deteriorating conveyance infrastructure. At the same time, the level of technical preparation and education of farmers is low, and Mexico's extension service has not efficiently upgraded the technical competence of farmers using irrigation.

"On Mexican irrigated farms, subsidies destroy incentives for water conservation and water is treated as essentially a 'free' good."

At the district or unit level, SARH's practice of leaving various projects incomplete and deferring maintenance on others undermines irrigation efficiency. So does the lack of connection between SARH's charge for water in irrigation districts and maintenance activities in the districts—a discrepancy that erodes farmers' confidence in SARH's rationale for assigning fees.

Finally, SARH's general failure to control groundwater exploitation is (with notable exceptions in the Northwest) leading to serious water depletion and contamination in several northern areas and permitting sea-water intrusion in coastal aquifers in the Northwest. If current trends continue, Mexican policy-makers face intractable problems of water scarcity and water quality in important population centers and agricultural areas—notably, in the Torreon-Lerma area (the Comarca Lagunera) and in Baja, California.

Clearly, a wide range of policies and management practices on the farm and in the irrigation districts and units are working against the Mexican government's efforts to increase agricultural productivity. These practices cannot continue if Mexico is to establish a sustainable agricultural economy. Fortunately, wide-ranging, cost-effective, and technically viable alternatives are available.

These alternatives for irrigated agriculture fall into two complementary categories: improved efficiency in water resource use, and strategic research and extension efforts. The water-management policies combat both surface and groundwater-use inefficiencies, primarily with infrastructure improvements and regulatory changes. The research and extension policies push for appropriate technology that all farmers can get and use.

Using Water More Efficiently

RECOMMENDATION 1: Focus SARH expenditures on project completion and project maintenance.

Given the high marginal costs of adding to the irrigation land base, the efficiency returns of upgrading Mexico's existing system would seem to outweigh the net returns that SARH can expect from constructing new projects. As noted by Crosson and his colleagues (1976), SARH must change its personnel and mental state to reflect the fact that SARH's "construction era"—the 1940s through the 1960s—must now give way to a "management era." Until it does, little progress can be expected in improving system efficiency given the sorry state of repair of many of Mexico's irrigation districts.

Regardless of which policy path Mexico takes to boost agricultural productivity, system maintenance must figure centrally and SARH must complete works at established irrigation districts and units. Completing deferred systems would both improve a project's overall efficiency and inspire participants' confidence that, for instance, proposed fees and charges for water-system maintenance are reasonable.

RECOMMENDATION 2: Provide incentives for farmers to improve the efficiency of water use.

If water is a "free" good, it will be used accordingly. Conversely, farmers will conserve the resource only if they have incentives to do so. One such incentive is to price the resource at its "scarcity" value or, at a minimum, at its full costs. Given economic conditions in Mexican agriculture today, adding to farmers' costs now may be politically infeasible; at any rate, any water-pricing changes would require a comprehensive review of all of Mexico's agricultural-pricing strategies. An alternative that could accomplish the same goal is to replace the stick with a carrot by offering rewards for improved efficiency. Thus, awards of either money or recognition (via, as an example, "farmer of the month" programs, dinner with State governors, and the like) for measured gains in efficiency might be implemented. Such efforts would raise resource-management standards and expectations.

RECOMMENDATION 3: Reduce the rate of groundwater mining.

Most mining of groundwater reserves in Mexico is by farmers—usually, private (non-*ejidal)* farmers—eager to supplement scarce allocations of surface waters. Groundwater mining, as practiced in Mexico, worsens income distribution, ultimately exhausts the resource, and increases political pressure to make costly interbasin transfers of water as "rescue operations" (Howe and Easter, 1971; and Larenas Davila, 1972).

"Groundwater resources are valuable only if used, but the issue now is use at rates far higher than those that optimally serve the region or the country's best interests."

Of course, groundwater resources are valuable only if used, but the issue now is use at rates far higher than those that optimally serve the region or the country's best interests. Such *was* the case in the Costa de Hermosillo before around 1975; such *is* the case in the Comarca Lagunera region today. (Such is also the case in large parts of the Texas high plains and other parts of the western United States.)

Two major policy alternatives may be considered for dealing with the common property resource. In the first, farmers are assigned property rights—which are marketable—to some portion of water in the aquifer (based, for example, on the amount of land overlying the aquifer that the farmer owns) (T. Anderson, 1984). Presumably, users will trade marketable water rights so as to optimize both the timing and rate of aquifer exploitation. However, this option may well clash with the Mexican Constitution, which makes all water resources state property. Under the second policy option, the state determines an "optimal" rate for exploiting the groundwater stock and then assigns farmers appropriate pumping rates—either setting water prices that induce users to pump at socially desirable rates or simply mandating pumping rates for farmers. This alternative, used in the Costa de Hermosillo in the late 1970s (see Cummings, 1974), entails enforcement problems, but can work if farmers can be prevented from pumping at rates that exceed allotted quotas.

Improving Research, Extension, and Technology

RECOMMENDATION 4: Improve the quality of agents in the extension service.

One immediately apparent policy option for inducing improvements in production technologies in Mexican irrigation is to substantively upgrade the extension service. Interestingly, in 1982 SARH prepared a proposal for accomplishing this end: each district or unit would have two professional extension staff, two vehicles, and a budget for trials and training, while every *three* districts would additionally share four researchers, technical assistants, field assistants, and a budget for vehicles, trials, and equipment.

The goals of the program proposed (for the northwest region) would be to increase yields for maize, beans, and cotton by 30 percent, for sorghum and alfalfa by 20 percent, for sesame, soybeans, and safflower by 15 percent, and for rice, wheat, and sugar cane by 10 percent. The training of extension agents would be upgraded; and salaries for agents would rise from US$169/month to $1,131/month (1985 dollars; SARH 1982, p. 32). For the northwest region, this proposed extension program would cost some $17 billion pesos (1982 pesos) over six years. But the program would yield an estimated benefit-cost ratio of 1.81 in the initial year, rising to a ratio of 5 at the end of six years. Similar proposals have been prepared for irrigation entities elsewhere in Mexico (*Programa,* pp. 31–40).

RECOMMENDATION 5: Make agricultural research more relevant.

Given the already scarce financial resources available for research in Mexico, authorities must carefully weigh priorities before assigning resources among competing research projects. Equally obvious, the most important criterion in assigning such priorities is the relevance of research. Technologies that are broadly applicable throughout Mexico's irrigated agriculture, as well as technologies that substantial numbers of farmers will adopt, should receive highest priority.

* * * * * * * *

Finally, it must be emphasized that while implementing these recommendations could improve productivity and yields from existing irrigated lands, doing so may not by itself be enough to meet the total demand for food and other agricultural products considering how rapid population growth is now. A much greater investment by the Mexican government in the rainfed sector—for research and extension, improved seed and fertilizer, afforestation, and soil conservation—is also needed to meet the goal of food security and to combat malnutrition.

Ronald G. Cummings, Chairman, Department of Economics, University of New Mexico. **Victor Brajer,** Associate Professor of Economics, California State University at Fullerton, California. **James W. McFarland,** Dean, College of Business Administration, Tulane University. **Jose Trava,** Professor, *El Colegio de la Frontera Norte,* Tijuana, Mexico. **Mohamed T. El-Ashry,** Vice President for Research and Policy Affairs, World Resources Institute, Washington, D.C.

NOTES

1. Suggestions of lower growth rates between 2.15 and 2.30 for the period 1980–2000 have been given by G. Aguilera, 1985, p. 87.

2. Rural population for 1985 estimated at 30.5 million people, expected to rise to 33 million by 1988; Secretaria de Agricultura y Recursos Hidraulicos (SARH), 1985, pp. 30 and 98.

3. Most of the following discussion of geographic conditions is based on data found in Venezian and Gamble, 1969, Chapters 1 and 2.

4. For more detailed discussions of the Mexican financial "crisis," and more particularly, of its problem with inflation, see M. Gonzalez, 1983; Davila, et al, 1984; Salas and Ize, 1984.

5. The above and following discussions draw upon analyses given in Venezian and Gamble, 1969, Chapters 1 and 2; and in Yates, 1981, Chapter 1.

6. See Cummings, 1972 and Venezian and Gamble, 1969, pp. 131–32. An exception to this observation is seen in SARH activity in the Costa de Hermosillo and in the Lerma Valley in the State of Mexico, where well fields supply potable water for Mexico City.

7. Information obtained by authors from the SARH office in Lerma Durango, January 1986; *see also* V. Sanchez as well as Flores Lui in *Consejo Nacional,* 1986, pp. 13 and 46.

REFERENCES

Aguilera G., Manuel, "La Agricultura Mexicana Hacia El Año 2000: Opciones, Limites Y Desafios," *Investigacion Economica,* 171, 85–129, January–March 1985.

Alba, Francisco, "Mexico's Northern Border: A Framework of Reference," *Natural Resources Journal,* 22(4), 749–764, October, 1982.

Anderson, R.J. and S. Fernandez, "An Illustration of the Potential Effects of Plastics Mulch Technology on Farm Income and Consumption of Plastics in Mexico."

Anderson T., "The Market Alternative for Hawaiian Water," *Natural Resources Journal* 25, 5, 893-910, October, 1985.

Andrade, M. Carlos, "La Tenencia de la Tierra y la Evolucion Politica de Mexico," *Revista de Economia,* Vol. XXIV, No. 10, Mexico, D.F., 1961.

Banco de Mexico, *Indices de Precios,* Subdireccion de Investigacion Economica, Cuaderno Mensual, 79, Mexico, D.F., November, 1984.

Banco de Mexico, "Datos Del Modelo Macroeconomico Para Mexico, 1960–83," mimeo, data made available by Dr. Javier Salas, January 1986.

Barkin, David, "Mexico: Tres Crisis Alimentarias," *Nexos,* No. 13, Mexico, D.F., 1983.

Brajer, Victor and Ronald G. Cummings, "Water Subsidies in Mexico's Irrigated Agricultural Sector," mimeo, World Resources Institute, Washington, D.C., 1986.

Brajer, Victor and Ronald G. Cummings, "The Use of Plastic Mulch in Northern Mexico," mimeo, Department of Economics, University of New Mexico, 1985.

Chonchol, Jacques, "La Reforma Agraria y El Desarrollo Rural como Estrategia de un Nuevo Orden Economico Internacional," *El Trimestre Economico,* XLIX(2), 194, 253–64, April–June 1982.

Clarkson, V.A. 1960, "Effect of black polyethylene mulch on soil and microclimate temperature and nitrate level. . .," *Agronomics Journal,* 52:307–309.

Coffey, David L. "Making the Most of Mulch," *American Vegetable Grower,* April 1984.

Consejo Nacional de Ciencia y Tecnologia (joint with: Ingenieros Agronomos del Tecnologico de Monterrey, A.C.) "Memorias del II Ciclo Internacional de Conferencias, Approvechamiento de Aguas Subterraneas en la Agricultura," meetings held in Torreon, Coahuila, Mexico (March 13–14, 1985, January, 1986).

Corden, W.M., "Booming Sector and Dutch Disease Economics: Survey and Consolidation," *Oxford Economic Papers,* 3(3), 359–80, November, 1984.

Crosson, P.R. et al. *Selected Water Management Issues In Latin American Agriculture,* Johns Hopkins Press, Baltimore, 1978.

Cummings, Ronald G., *Water Resource Management In Northern Mexico,* Johns Hopkins Press, Baltimore, 1972.

Cummings, Ronald G. *Interbasin Water Transfers: A Case Study In Mexico,* Johns Hopkins Press, Baltimore, 1974.

Cummings, Ronald G., "Some Extensions of the Economic Theory of Exhaustible Resources," *Western Economic Journal,* 7(3), 201–210, 1968.

Cummings, Ronald G. y Victor Brajer, "Seleccion de la Superficia Optima para el Cubrimiento con Tela de Plastico: Un Caso de Estudio," *El Trimestre Economico*, LVI(3), 223, July–September 1989.

Davila, Jose, Alain Ize and Jose Morales, "Fuentes del Proceso Inflacionario en Mexico: Analisis de Causalidad," pp. 57–70 in Ize Alain and Gabriel Vera (Editors), *La Inflacion En Mexico: Ensayos,* El Colegio de Mexico, Mexico, D.F., 1984.

Davila Jimenez, Gerardo, "La Politica de Precios Y Subsidios," pp. 228–276 in Gonzalez M. Hector E. (Editor), *El Sistema Economico Mexicano,* La Red de Jonas (Publisher), Mexico, D.F. Second Edition, 1983.

Escamilla, J. and N. Pinon, "Estudios de los Acusteros de la Frontera Norte," Comision del Plan Nacional Hidraulico, SARH, Mexico, DF, December, 1983.

Gonzalez, R. Fernando, "El Aspecto Agrario de la Revolucion Mexicana," *Problemas Agricolas y Industriales de Mexico,* V, 3, Mexico, D.F., 1953.

Harris, R.E., "Polyethylene Covers and Mulches for Corn and Beans Production in Northern Regions," *Proceedings of the American Society of Horticulture Sciences,* 87:288–94, 1965.

Hopen, H.J. and N.F. Oebker, "Vegetable crop responses to synthetic mulches," an annotated bibliography. Illinois Agricultural Experiment Station Special Publication 42, 1976.

Howe, Charles W. and William K. Easter, *Interbasin Transfers of Water: Economic Issues and Impacts,* Johns Hopkins University Press, Baltimore, 1971.

Huizer, Garit, "La Lucha Campesino En Mexico," Mimeo, Centro de Investigaciones Agrarias, Mexico, D.F., 1970.

Instituto Nacional de Estadistica, Geografia E Informatica, *Sistema de Cuentas Nacionales de Mexico: Principales Variables Macroeconomicas Periodo 1970–1982,* Mexico, D.F., 1983.

Instituto Nacional de Estadistica, Geografia E Informatica, *Sistema de Cuentas Nacionales de Mexico: Estimacion Preliminar 1984,* Mexico, D.F., January, 1985.

Larenas Davila, Luis, *Aspectos Economicos Sobre La Transferencia De Agua Entre Cuencas,* unpublished M.A. thesis, Centro de Economica Agricola, Escuela Nacional de Agricultura, Chapingo, Mexico, 1972.

Lustig, Nora (Editor), *Panorama Y Perspectivas de la Economia Mexicana,* Colegio de Mexico, Mexico, D.F., 1980.

Merkley, Gary, and Jack Keller, "Irrigation Efficiency," *Memorias del II Ciclo Internacional de Conferencias: Aprovechamiento de Aguas Subterraneas en la Agricultura,* Ingenieros Agronomos del Tecnologico de Monterrey, A.C., in conjunction with Consejo Nacional de Ciencia Y Tecnologia, Torreon, Coah., Mexico, pp. 121–131, March 13–14, 1985.

Moore, Ronald L., "Eficiencia de Pozos de Agua," *Memorias del II Ciclo Internacional de Conferencias: Aprovechamiento de Aguas Subterraneas en la Agricultura,* Ingenieros Agronomos del Tecnologico de Monterrey, A.C., in conjunction with Consejo Nacional de Ciencia Y Tecnologia, Torreon, Coah., Mexico, pp. 121–131, March 13–14, 1985.

Munoz, Carlos, "La Tenencia de la Tierra y la Evolucion Politica de la Tierra," *Revista de Economia,* Vol. XXIV, No. 11, Mexico, D.F., 1961.

Nacional Financiera, "El Mercado de Valores" Ano XLV, Numbers 51 and 52, December, Mexico, D.F., 1985.

New Mexico State University, *New Mexico Agricultural Statistics 1982,* Las Cruces, NM, 1982.

Nicholas, George. "Plastic Mulching" *Farmer's Weekly,* June 10, 1983.

Nunn, Susan C., "The Political Economy of Institutional Change: A Distribution Criterion for Acceptance of Groundwater Rules," *Natural Resources Journal,* 25, 4, 867–892, October, 1985.

Oyarzabal, Francisco, "Comentarios a los Leyes e Instituciones que Reglamentan las Aguas Superficiales de Mexico," *Natural Resources Journal,* 22(4), 999–1006, October, 1982.

Palacios, Velez, Enrique, "Costo del Agua de Bombeo en la Region Lagunera," *Memorias del II Ciclo Internacional de Conferencias: Aprovechamiento de Aguas Subterraneas en la Agricultura,* Ingenieros Agronomos del Tecnologico de Monterrey, A.C., in conjunction with Consejo Nacional de Ciencia Y Tecnologia, Torreon, Coah., Mexico, pp. 33–68, March 13–March 14, 1985.

Rincon Valdes, Carlos A., "Aguas Subterraneas en la Region de Cuidad Juarez, Mexico," *Natural Resources Journal,* 22(4), 939–942, October, 1982.

Ruiz Duran, Clemente, "La Politica Monetaria Y Crediticia 1970–1981," pp. 69–99 in Violante, Alejandro and Roberto Davila (Editors), *Mexico: Una Economia En Transicion: Volumen I,* Editorial Limusa (Publisher), Mexico, D.F., 1984.

Salas, Javier and Alain Ize, "Dinero, Precios Y Producto: Un Analisis de Autorregresion Vectorial Para Mexico," pp. 71–86 in Ize, Alain and Gabriel Vera (Editors), *La Inflacion en Mexico: Ensayos,* El Colegio de Mexico, D.F. 1984.

Scott, Anthony, *Natural Resources: The Economics of Conservation,* McClelland and Stewart, Ltd., Carleton Library No. 68, Ottawa, 1973.

Secretaria de Agricultura y Recursos Hidraulicos, *Plan Nacional Hidraulico,* Mexico D.F., 1981 (A). *Produccion Agricola Nacional, 1981 Anuario Estadistico,* Direccion General de Economia Agricola, Mexico, D.F. 1981 (B).

____, "Informes de Riego, DGDUR, Mexico, D.F., 1981 (C).

____, "Estudio Para el Mejoramiento de la Produccion: Consideraciones Preliminares," Mexico, D.F., October, 1982.

____, *Econotecnia Agricola: Consumos Aparentes de Productos Agricolas 1925–1982,* Vol. VII, Num. 9, Mexico, D.F., September, 1983.

____, *Catologo De Cuotas Por Servicio de Riego,* Ciclo Agricola 1983–1984, Officina de Cuotas, Mexico, D.F., 1984 (A).

____, *Rentabilidad Y Dinamica de las Explotaciones Agricolas de Maiz en la Region de Huamantla, Tlaxcala,* Mexico, D.F., February, 1984 (B).

____, *Informe Estadistico,* No. 136, Mexico, D.F., April, 1984 (C).

____, *Acciones Instrumentadas en el Marco del Programa Nacional de Desarrollo Rural Integral 1983–1984,* Mexico, D.F., May 16, 1985 (A).

____, *Programa Nacional De Desarrollo Rural Integral 1985–1988, Poder Ejecutivo Federal,* Mexico, D.F., Second Edition, June, 1985 (B).

Senior, Clarence, *Land Reform and Democracy,* Greenwood Press, Westport, CT, 1974.

Spice, H.R. 1977. *Polyethylene Film in Horticulture.* Latimer Trend and Co. Ltd., Plymouth, London.

Trava, Jose, "Efficiencias en Distritos de Rieso del Noroeste" Centro de Estudios Fronterizos del Norte de Mexico (CFSDN) Mexicali, 1985.

____, "Tables Showing Land Quality and Aquifer Levels in Selected Irrigation Areas," CFSDN, Mexicali, 1987.

____, "Servicio De Extension Agricola a Nivel Nacional," mimeo, El Colegio de la Frontera Norte, Tijuana, B.C. July 21, 1986.

Trujillo, P.M. and J.N. Corgan, "Plastic Mulches, Hot Kaps, and Transplants for Vegetables in Northern New Mexico," Bulletin 682, New Mexico State University Agricultural Experiment Station, May 1981.

U.S. Department of Commerce, *Statistical Abstract of the United States 1986,* Bureau of the Census, Washington, D.C. 1986.

Venezian, E.L. and W.K. Gamble, *The Agricultural Development of Mexico,* New York: Praeger Publishers, 1969.

U.S. Department of Agriculture, *Handbook For Water Harvesting,* Agriculture Handbook No. 600, Washington, D.C., June, 1983.

Wall Street Journal, 'Nation in Jeopardy," series September 26–October 15, 1985.

____, "Mexico Peso Slumps on Consumer Price Increase," p. 29, January 13, 1986.

____, "Mexico and Venezuela Slash Oil Prices, Seek Meeting to Discuss Region's Debt," p. 3, February 3, 1986.

____, Mexico Tells Creditors It Will Need to Borrow $9 Billion This Year," p. 30, February 5, 1986.

World Bank, "Mexico Irrigation Subsector Survey—First Stage, Improvement Of Operating Efficiencies In Existing Irrigation Systems," Washington, D.C., June 20, 1983.

Yates, P. Lamartine, *Mexico's Agricultural Dilemma,* Tucson, Arizona: The University of Arizona Press, 1981.

Zamorano Ulloa, Jose, "La Politica Alimentaria en Mexico 1980–85: Balance Y Perspectivas," Colegio de Mexico, Mexico, D.F. 1985.

APPENDIX

Reintroducing Native Vegetation: An Alternative to Land Abandonment

by Manuel Puebla* and Mohamed T. El-Ashry

Once the new technologies or techniques so essential to improving irrigated agriculture become available, most farmers either adopt them or abandon farming. Converting is obviously preferable, but not all farmers have the expertise or funds needed to finance expensive new technology. Many have no choice but to abandon the land. Indeed, 13 percent of the three irrigation districts' productive irrigated lands have already been abandoned.

Arguably, making do with less water need not be a big problem since 40 percent of Mexico's irrigation is in Sonora and Sinaloa, where farmers have comparative technological, educational, and economic advantages over most other Mexican farmers. But these are advantages that only private farmers with large holdings enjoy. Small private farmers have just as much trouble obtaining credit as *ejidal* farmers and even more problems getting technical advice since most extension agents help only *ejidal* farmers (Yates, 1981, p. 123). In any case, a quarter of these districts' irrigation lands are under *ejidal* and community ownership. Few such operations are credit-worthy and, despite limited technical assistance, most lack the flexibility to adapt.

Abandonment hurts the farmer, the community, and the country. Farmers left demoralized and landless have only two options: either they remain in the area, thus adding to pressures on the impoverished rural economy, or they join the hoards of unemployed in overpopulated cities. The land they abandoned, once stripped of the native soil-stabilizing vegetation and encrusted with salts, is open to severe wind and water erosion. Deposits of the poor soil from these barren lands may also jeopardize the productivity of adjacent cultivated fields.

One solution to this waste of human and natural resources is to reintroduce relatively salt- and drought-resistant native vegetation. Studies have

*Director, Centro de Investigacion y Desarrollo de los Recursos Naturales de Sonora (CIDESON).

shown that many native arid-land plants in northern Mexico have commercial value, including *jojoba* (an oil base), *guayule* (a source of rubber), and *candelilla* (a source of wax) (NAS, 1975a). In fact, some of these plants are currently being promoted in Sonora and other parts of the country, while mesquite (a multipurpose plant) and others have been used to halt desertification in other countries (NAS, 1975b).

The transition to a commercial native-crop economy, however, will not happen by itself. Reintroducing the plants involves significant labor, capital investment, and some risk, and many of these plants must be established for two to five years before a commercially valuable crop can be harvested. Subsistence farmers are not likely to assume large costs or risks alone, so the Mexican government has a great role to play. It could provide economic incentives for growing and marketing the plants, as well as for research and extension services.

This appendix provides a brief discussion of the potential for reintroducing commercially valuable native vegetation so as to minimize the adverse impacts—including land abandonment—of potentially useful water-management policies in Mexico's arid and semi-arid regions. Brief descriptions of the plants, their uses, and markets are given. Plants with potential markets are identified. The benefits and risks involved are presented very broadly. Recommendations cover government's role and the government incentives that could be offered to support such an initiative.

Native Plants of Commercial Value

That many plants native to the Sonoran Desert and Mexico's northwest region have commercial value is not a recent discovery. At the turn of the century, guayule rubber extraction facilities were built in Mexico (NAS, 1975b). Jojoba oil has been studied and used since the 1950s in Mexico (NAS, 1975a). Only in recent years, however, have many desert plants become economically attractive. A global awareness that petroleum and groundwater are finite and that the earth's biodiversity must be maintained has prompted increased activity in developing sustainable management plans for nonrenewable resources and product substitutes.

Six plant types are covered here as examples of the vegetation that could be commercially cultivated. Other plants, such as cactus, canaigre, guar, nopal, saltbush, senna, and tepary beans, are not described here, but some are currently grown in northern Mexico for their commercial value, while others are under investigation by the *Centro de Investigacion y Desarrollo de los Recursos Naturales de Sonora* (Vietmeyer, 1986).

In general, these six plants, deep-rooted and salt tolerant, demand little water. All make fine hedges against desertification, and some of the nitrogen-fixing varieties may even improve the quality of the original soil. Many require little or no fertilizer or maintenance and can resist many pests and diseases. Seed propagation and harvesting processes (using conventional or slightly adapted machinery) can be mechanized for some. These crops can be used to great benefit even where groundwater is not in short supply, and intercropping some of these plants with other crops often increases the yields of both (Nair, 1980).

JOJOBA
Simmondsia chinensis

MARKETABLE PROPERTIES

—Possesses many of the same properties as sperm whale oil.
—Viscosity constant at any temperature.
—Does not absorb oxygen or become rancid or spoil at high temperatures.
—Contains no fats or other components that require refining.

USES

Cosmetics, candles, food for people and animals, leather softener, pharmaceuticals, lubricants, and solvents.

MARKETS

Used in American, Japanese, and European cosmetics industry and has potential markets as lubricants, solvents and pharmaceuticals (Menanteau, 1988).

PHYSIOLOGY AND COMPOSITION

—Requires little water during flowering and seed formation.

—Long central root allows tapping of water at depth.
—Takes five years to produce a harvest of economic value.
—Lives up to 200 years.
—Seeds contain 50 percent oil.
—Oil has 36 percent protein.

DISTRIBUTION/YIELD
—Native to the Sonoran Desert in Mexico, California, and Arizona.
—Grows at elevations of 2000 to 4000 feet.
—Yield of total acreage of wild plants is over one million pounds annually.
—Yields of cultivated plants are from 1.5 to 12 pounds of seeds per plant.

GUAYULE
Parthenium argentatum

MARKETABLE PROPERTIES
—When purified, indistinguishable from natural Hevea rubber.

USES
Rubber products—principally tires.

MARKETS
Currently a minor rubber source. Pilot rubber-processing plant established in Mexico in mid-1970s. Amerind Agrotech Laboratories of the Gila River Indian Community in Sacaton, Arizona (a joint effort of the U.S. Departments of Agriculture and Defense) will be producing rubber by 1990s.

PHYSIOLOGY AND COMPOSITION
—Slow-growing perennial.
—Needs three years for harvestable crop.
—Lives about 50 years.

DISTRIBUTION AND YIELD
—Native to northern Mexico and the southwestern United States.
—Rubber contained throughout plant.
—Yields from wild plants 12 percent, cultivated species up to 20 percent by dry weight (up to 800–900 kilograms of rubber per hectare for improved varieties of guayule).

CANDELILLA
Euphorbia antisyphilitica

MARKETABLE PROPERTIES
—Amorphous solid yellow wax.

USES
Substitute for beewax and carnauba wax coatings and polishes, hardener for softer waxes, chewing gum, and other products.

MARKETS
Mexican government subsidizes industry for wild candelilla in rural communities. Five million kilograms exported annually to United States by Mexico at $1.50 per kg at 1975 dollars.

PHYSIOLOGY AND COMPOSITION
—Leafless desert scrub.
—Requires little rainfall.
—Needs two to five years to produce commercially adequate amount of wax.

DISTRIBUTION AND YIELD
—Native to the Coahuilan desert in Mexico and in southern New Mexico and Texas.
—Two-to-five percent wax yield using simple processing methods.

MESQUITE
Prosopis spp

MARKETABLE PROPERTIES
—Edible fruit and leaves for people and livestock.
—Heavy and hard wood.
—Medicinal properties.
—Nitrogen fixer.

USES
Food products, forage for livestock, fuelwood, building material, medical and cosmetic products, and soil nutrient.

PHYSIOLOGY AND COMPOSITION
—Medium-size shrubs or spreading, short-trunked trees.
—Can tolerate and grow rapidly in barren wastelands if sufficient subterranean water exists.
—Easily withstands protracted droughts.

—Pods have a feed value roughly comparable to that of barley or corn (34–39 percent protein; 7–8 percent oil).

DISTRIBUTION AND YIELD
—Native to southern North America—currently grown in Mexico, 1000+ hectares in Peru and Chile and 100,000 hectares in Brazil for commercial uses.
—Yields 2 meter by 10 centimeter poles within two years.
—Fruit yields range from 5 to 100 kilograms per tree or 50 to 2,000 kilograms per hectare.

AMARANTH
Amaranthus hypochondriacus

MARKETABLE PROPERTIES
—High lysine content in protein, similar to that of soymeal.

USES
Food, potherbs, and ornamentals.

PHYSIOLOGY AND COMPOSITION
—Cereal-like plant with large sorghum-like seeds.
—Requires same growing area as maize plant.
—Requires good well-tilled soil and moderate rainfall.
—26 percent germ and bran and 74 percent flour in a seed; 20 percent oil in germ (NAS, 1984).

DISTRIBUTION AND YIELD
—Native to South and Central America and Mexico.
—Seeds yield over 60 percent starch, 30 percent protein (NAS, 1984) and 6 percent lysine per gram protein.
—Some harvests have produced 1 ton per hectare (NAS, 1975b).

BUFFALO GOURD
Cucurbita foetidissima

MARKETABLE PROPERTIES
—Polyunsaturated, protein-rich oil.
—Starch from root similar to cassava root (Hinman, 1986).
—Soapy extract from fruit and vine.

USES
Edible oil for human and livestock consumption, sweetening agent, cleaning solution, liquid fuel, and cattle feed.

PHYSIOLOGY AND COMPOSITION
—Perennial vine.
—Large roots that penetrate ground for water.
—Seeds contain 30 to 35 percent protein and up to 34 percent oil.
—Lives up to 40 years.

DISTRIBUTION AND YIELD
—Grows wild in the deserts of Mexico and the southwestern United States, but is not commercially cultivated anywhere.
—Yields 60 fruits per plant with 12 grams of seeds per fruit (seed yield comparable to soybean and peanuts).
—Roots weigh up to 30 pounds in 2 years and yield up to 8 tons per hectare (Hinman, 1986).

Recommendations

Since neither the rural nor urban economies can absorb dislocated farmers, and since abandoning lands would incur significant economic and ecological costs to the country, introducing native plants that could keep the farmers employed on their farms is an attractive and relatively low-cost alternative. As shown in the previous section, many plants native to northern Mexico are being developed into commercial crops. Besides their export value, many are sources of nutrition for villagers and their livestock—top priorities for the Mexican government.

The other potential benefits of these underutilized plants are manyfold: they yield a variety of products with industrial applications, and they can revitalize nutrient-depleted soils, halt desertification, and reestablish the shrinking genetic diversity of food crops. Dr. N.D. Vietmeyer notes: "Throughout history mankind has used some 3,000 plant species for food, but over the centuries the tendency has been to concentrate on fewer and fewer. Today, most of the world's food comes from a mere 20 or so species." (1986, p. 1379).

These benefits notwithstanding, most farmers lack the resources and expertise to undertake a revegetation project on their own and the Mexican government does not possess endless resources for every worthwhile cause. However, the government does have experience in developing some of these native crops and has laid the groundwork for future research and expansion.

Of course, some uncertainties are involved in producing these plants. Harvestable yields may take over two to five years to produce, yields of cultivated plants are still quite variable, and the proper agronomic techniques for optimal yields require more research. Seed propagation, maintenance, and harvesting for some plants remains laborious and painstaking because a suitable mechanization has not been developed. Few statistics exist on the marketing of many of these native plants, and world markets often discourage small-scale product development. For these reasons, many aspects of the commercialization of these plants still need to be investigated—among them, methods of ensuring consistent and abundant yields from cultivated areas; processing methods that will capitalize on new and optimal properties from the plants; various product applications for new and existing markets; and the economics of distribution, processing, manufacturing, advertising, and sales of the products.

Similarly, the training and promotional needs associated with introducing new crops could be included as part of the extension training program improvements of the Mexican Secretariat of Agriculture and Water Resources (SARH). Agents could be instructed in the proper agronomic and technological methods for cultivating native plants. Research at CIDESON has indicated that native vegetation grown on damaged or marginal lands can, with the proper integrated mix of sustainable agronomy and technology out-produce, at a better rate of return, unintegrated farming approaches on better soils with the same plants. *(See Tables A-1, A-2, and A-3.)* The integrative approach incorporates seed treatment to improve germination, the application of steroidal phytohormones to increase yields, solar and wind power for water pumping, and various irrigation technologies for optimizing water uptake and conservation. Extension agents would also play a crucial role in promoting the concept and its economic feasibility among farmers.

Table A-1. Jojoba Investment Rate of Return (IRR) ($ = Mexican Peso)

Year	Cash Flow Control	Cash Flow Integrated Approach
0	$-33113	$-36415
1	-1190	-68
2	-304	1439
3	878	3448
4	3832	8469
5	7524	14746
6	17124	31066
7	27462	48641
8	37801	66215
9	48139	83792
10	57001	98857
	IRR = 24.87909	IRR = 36.98486

Table A-2. Jojoba Seed Production (Kg/ha)

Year	Control	Integrated Approach
0	8.0	13.6
1	60.0	102.0
2	120.0	204.0
3	200.0	340.0
4	400.0	680.0
5	650.0	1105.0
6	1300.0	2210.0
7	2000.0	3400.0
8	2700.0	4590.0
9	3400.0	5780.0
10	4000.0	6800.0

Finally, legislation and funding for economic incentives and risk-sharing measures directed at farmers are essential to successful program implementation. Before the government establishes a new subsidy or loan program for native plants, the

Table A-3. Jojoba Seed Production Value (U.S. Dollars)

Year	Control	Integrated Approach
0	$ 118.0	$ 201.0
1	886.0	1506.0
2	1772.0	3013.0
3	2954.0	5022.0
4	5908.0	10043.0
5	9600.0	16320.0
6	19200.0	32640.0
7	29538.0	50215.0
8	39877.0	67791.0
9	50215.0	85366.0
10	59077.0	100431.0

agricultural pricing system should be examined for distortions that could be corrected to provide the needed incentives. Lessons may be learned by looking at the incentive programs utilized for initiating the current small-scale native plant projects. Any subsidies considered should be tailored specifically to benefit the desired farmer population or water-management goals. Two examples of such targeted subsidies are low-or no-interest loans for the special machinery and farm inputs needed to cultivate the native plants being promoted and financial subsidies to cover the inputs and sales of these plants and their products. Assistance programs should be only short-term, providing support for the first few uncertain growing years. The cost of such subsidies could be covered by the revenues received from groundwater pumping or increased irrigation water rates.

In conclusion, the Mexican government has historically made substantial investments in developing a viable and productive irrigation sector, and it is progressing with its water-conservation efforts. There is significant reason to protect these investments. Nonetheless, the finite and vulnerable nature of soil and groundwater resources in northwest Mexico threaten the long-term success of the government program. Fortunately, ecologically and economically sound alternatives to traditional crops with high input demands do exist. Farm communities can return their damaged lands with little or no irrigation to a productive state by introducing commercially valuable native vegetation *if* government provides the necessary encouragement, guidance, and financial support.

REFERENCES

Hinman, C.W. "Potential New Crops." *Scientific American*. Vol. 255, No. 1, 1986.

Menanteau, Philippe. "Jojoba, What About It?," *African Environment*. pp. 161–165, 1988.

Nair, P.K.R. *Agroforestry Species: A Crop Sheets Manual*. International Council for Research in Agroforestry, Nairobi, Kenya, 1980.

National Academy of Sciences. *Amaranth: Modern Prospects for an Ancient Crop*. Washington, D.C., 1984.

____, *Products from Jojoba: A Promising New Crop for Arid Lands*. 1975 a.

____, *Unexploited Tropical Plants with Promising Economic Value*. 1975 b.

Vietmeyer, N.D. "Lesser Known Plants of Potential Use in Agriculture and Forestry." *Science*. Vol. 232, pp. 1379, 1986.

World Resources Institute

1709 New York Avenue, N.W.
Washington, D.C. 20006, U.S.A.

The World Resources Institute (WRI) is a policy research center created in late 1982 to help governments, international organizations, and private business address a fundamental question: How can societies meet basic human needs and nurture economic growth without undermining the natural resources and environmental integrity on which life, economic vitality, and international security depend?

Two dominant concerns influence WRI's choice of projects and other activities:

The destructive effects of poor resource management on economic development and the alleviation of poverty in developing countries; and

The new generation of globally important environmental and resource problems that threaten the economic and environmental interests of the United States and other industrial countries and that have not been addressed with authority in their laws.

The Institute's current areas of policy research include tropical forests, biological diversity, sustainable agriculture, energy, climate change, atmospheric pollution, economic incentives for sustainable development, and resource and environmental information.

WRI's research is aimed at providing accurate information about global resources and population, identifying emerging issues, and developing politically and economically workable proposals.

In developing countries, WRI provides field services and technical program support for governments and non-governmental organizations trying to manage natural resources sustainably.

WRI's work is carried out by an interdisciplinary staff of scientists and experts augmented by a network of formal advisors, collaborators, and cooperating institutions in 50 countries.

WRI is funded by private foundations, United Nations and governmental agencies, corporations, and concerned individuals.